THE RACE FOR THE ATOMIC BOMB

THE RACE FOR THE ATOMIC BOMB

SCIENTISTS, SPIES AND SABOTEURS THE ALLIES' AND HITLER'S BATTLE FOR THE ULTIMATE WEAPON

NORMAN RIDLEY

FRONTLINE BOOKS

First published in Great Britain in 2023 by
FRONTLINE BOOKS
an imprint of Pen & Sword Books Ltd,
47 Church Street, Barnsley, S. Yorkshire, S70 2AS

ISBN: 978-1-39904-032-7

CIP data records for this title are available from the British Library

For more information on our books, please visit
www.frontline-books.com, email info@frontline-books.com
or write to us at the above address.

Printed and bound by CPI Group (UK) Ltd, Croydon, CR0 4YY
Typeset by Concept, Huddersfield, West Yorkshire
Pen & Sword Books Ltd incorporates the imprints of Pen & Sword
Archaeology, Atlas, Aviation, Battleground, Discovery,
Family History, History, Maritime, Military, Naval, Politics,
Social History, Transport, True Crime, Claymore Press,
Frontline Books, Praetorian Press,
Seaforth Publishing and White Owl

For a complete list of Pen and Sword titles please contact
PEN & SWORD LTD
47 Church Street, Barnsley, South Yorkshire, S70 2AS, England
E-mail: enquiries@pen-and-sword.co.uk

Or

PEN AND SWORD BOOKS
1950 Lawrence Rd, Havertown, PA 19083, USA
E-mail: Uspen-and-sword@casematepublishers.com

CONTENTS

LIST OF PLATES

'These processes are in fact similar to those which took place in the early stages of development of the universe and still go on in the turbulent and flaming interior of stars . . . A weapon of devastating power far beyond any previously possibilities and imagination will soon become available.'

Extracts from a letter written by Niels Bohr to
Winston Churchill, 22 May 1944

'Germany is attempting to build a plant like this. Whoever gets there first will win the war.'

Lieutenant-Colonel Franklin Matthias, Corps of Engineers Chief
at the Hanford Site during the Manhattan Project in a speech to
construction workers at White Bluffs, Spring 1944

PREFACE

On 19 December 1938, Otto Hahn, working at the Kaiser Wilhelm Institute for Chemistry in Berlin, conducted an experiment the results of which baffled him. It took his émigré collaborator Lise Meitner to explain that he had actually split an atom of uranium, which at the time seemed to defy all known laws of physics. At this time, Germany, along with France, was in the forefront of atomic research.

When France fell to the German Blitzkrieg in 1940, most of its nuclear scientists had fled to Britain. For their part, the Germans had weakened their own scientific ranks by hounding out many of their best scientists, who had fled persecution under the draconian Nazi racial laws. They still retained, however, possibly the ablest nuclear scientist of them all in Werner Heisenberg, who set about developing his own programme for nuclear power.

When Niels Bohr awakened America to the dangers of atomic weaponry of immense power it was the fear of Nazi Germany exploiting its pre-eminent position to forge ahead in pursuit of developing an atomic bomb that forced Britain and the US to pool their resources, create the Manhattan Project and race the Germans to the finishing line. It had been the British who had made the early breakthroughs in research but it soon became obvious that they lacked the resources that would be necessary to put the theory into practice, so they offered up all their findings to the industrial might of America.

Whilst there was a general consensus that the huge challenges of industrialising the science would push the reality of a usable bomb beyond the probable duration of the war, there lingered a suspicion that the Germans, with their vast industrial expertise and capabilities alongside their utilisation of massed slave labour, would defy the odds. The indications were that Heisenberg was leading Germany in a programme of rapid research and development and US President Franklin D. Roosevelt was not prepared to take the chance that he would succeed.

The Soviets at this time were heavily involved in their own existential battle with Germany and showed little enthusiasm for the sort of investment required to build their own bomb but with an eye to the future they established an extensive espionage network both in Britain and the US to try and keep abreast of developments there. The Allies also developed intelligence agencies especially designed to find out just how far the Germans had progressed along the road to an atom bomb and they organised missions to sabotage and degrade any facilities the Germans had established.

After D-Day, Allied scientists followed troops into occupied Europe and searched for evidence of the German bomb. Soviet forces were doing the same from the east and both sides were anxious to gather up as much equipment and as many German scientists as they could before the other side beat them to it.

INTRODUCTION

'Now I am become Death, the destroyer of worlds'

The *Bhagavad-Gita*

Sixty million years ago, an asteroid between 10 and 15km wide struck the Yucatan Peninsula in what is now Central America at a speed of around 30km per second and caused an explosion the consequences of which destroyed most of life on earth. The devastation wrought by the collision had resulted from a combination of the mass (amount of matter in the asteroid), which instantly vaporised in the collision, and its kinetic energy (energy of motion) that was released as it came to an abrupt halt, gouging out a crater 150km wide. It was this sudden catastrophic release of energy which caused the explosion that hurled billions of tons of debris from the crater into the atmosphere, darkening the skies for many years afterwards and raising the temperature of the planet. This was but a tiny manifestation of the way that energy was bound up in the universe and how it could be transformed from one kind to another with stunning effect.

Around the ninth century, during the late Tang Dynasty in China, scientists made the first discoveries of a different kind of energy transformation. They succeeded in releasing the energy that was tied up in the bonds that held chemical compounds together. Of course, chemical bonds are made and broken all the time in nature, which is most obvious in the way that food is digested to power the human body, but these are relatively slow processes and take place in a non-violent way. What the Chinese scientists had discovered was that certain chemical bonds could be broken instantaneously which resulted in a sudden release of energy. They experimented by mixing and heating certain materials which they found to 'fly and dance' and, if they went on to confine them while heating, a pressure would build up extremely rapidly causing the retaining vessel to explode. What happened was that the chemical bonds holding the original materials together were broken down, as the materials reacted with each other to produce new and different materials and were replaced by weaker bonds that bound the new materials. The difference in the total bond energies of the original materials and the total bond energies of the new materials was released in the form of kinetic energy which caused the explosion. Whilst the Chinese discovery of gunpowder completely transformed warfare, it had no direct impact on the everyday life of non-belligerents. This would come about several centuries later during the Industrial Revolution.

For around 60 million years at the end of the Carboniferous Period (359–229 million years ago) the power of sunlight was captured along with vast quantities of carbon dioxide to generate great forests that covered much of what is now the northern hemisphere. As the vegetation in these forests grew and regenerated over this enormous time span the plant debris accumulated, was deposited on the forest floor and slowly compressed as more and more built up above it. The result of this was the carboniferous coal seams that were later exploited to drive the Industrial Revolution. To watch a piece of coal burn and feel the heat from it is to experience the energy of the sun, that has been trapped inside it for over 300 million years, released from its bondage. Along with that came the release of the carbon dioxide sucked in from prehistoric atmospheres to burst free again with all the existential threat that poses now. Oil deposits had a different origin, but the story is very much the same for them also. Although the impact of coal and oil was different to that of gunpowder, the power of their material transformations was still of a chemical nature. The burning of coal and oil is nothing more than the breaking down of chemical bonds in complex hydrocarbon compounds by the application of heat into simpler ones such as carbon dioxide and water. Soon, however, an energy source would be discovered that heralded a new dimension to industrialisation, but it would also bring in with it the dark spectre of a destructive force that had the potential, at a stroke, to destroy all life on earth much as the Yucatan asteroid had almost done.

Albert Einstein theorised that almost boundless energy was trapped inside the very heart of matter, the atoms themselves that were the building blocks of all compounds in the universe. This was energy that had been there not for millions of years but from the very beginning of time. It was energy that had been captured and locked up in atomic nuclei that were forged in the heart of massive stars at intense temperatures and pressures and, in the case of the heavier elements, in supernovae explosions which are the most violent events in the whole universe. Only in these explosions, where conditions are even more extreme than those in the centre of giant stars, are the atoms of such as uranium, the heaviest naturally occurring element, created. The extreme energies of those events would become transformed into the very matter of atoms as they were created and under normal conditions would go on to remain fixed there. Einstein gave the mathematical proof showing the relationship between the mass of an atom and the equivalent amount of energy bound up in that mass but, as a theoretical physicist, he gave little consideration to the practical applications of his results. While the chemical energy in coal and oil was the entrapment by plants of sunlight reaching the earth across space, this newly discovered atomic energy at the heart of matter had harnessed the fundamental power of the universe that been manifest in the raging furnaces of stars and the cataclysmic violence of exploding supernovae. The fission of 1kg of uranium-235, for example, produces about 2.5 million times as much energy as is produced by burning 1kg of coal. The very forces that had created the building blocks of life itself now threatened to break free of their shackles with the potential to finish the job that the Yucatan asteroid had only narrowly failed to. When scientists 20 years after Einstein's discovery began to unlock this power by physically splitting atomic nuclei the new Atomic Age was born.

Chapter 1

THE STRUCTURE OF MATTER

> 'It is impossible for atoms to come into existence out of nothing or pass away into nothing they are indestructible and eternal.'
>
> Epicurus (341–270 BC)

When trying to understand how scientists developed the ideas that led to the atomic bomb it is helpful to have some idea of the structure of atoms. It is obviously a vastly more complicated subject than could be covered in this book, but a simplistic overview is possible in which the basic elements of the theory allow the reader to appreciate more fully just how complicated it was, why it took such a monumental effort to build the bomb and why it was so devastating in its effects.

In 1903, Ernest Rutherford had shown that the energy contained in an atomic reaction was perhaps as much as a million times greater than the energy contained in a chemical reaction of the same mass. His colleague Frederick Soddy wrote 'there is associated with the internal structure of the atom an enormous store of energy which, in the majority of cases, remains latent and unknowable'.[1] In 1911, the two men explored the structure of the atom and showed that it had a nucleus, in which 99 per cent of the mass was concentrated, around which a cloud of electrons orbited. With the exception of the simplest form of the element hydrogen, the nucleus of every element is made up of two different components: protons and neutrons. The number of protons in the nucleus defines the element and gives its atomic number. Protons have a positive charge and electrons have a negative charge and each element has in its orbit the required number of electrons to balance the charge of the protons. Some elements, however, may vary in the number of neutrons their nucleus contains and in which case they are said to have isotopes. Some isotopes have unstable nuclei that naturally and slowly release excess energy by spontaneously emitting alpha, beta and extremely hazardous high-energy gamma rays.

Neutrons, discovered by James Chadwick in 1932, are particles formed when a positively-charged proton and a negatively-charged electron are joined together and so have no electrical charge at all. As an example, the simplest element is hydrogen which has only a single proton in its nucleus, hence atomic number 1 and a single electron orbiting it. Its nucleus may also contain one or two neutrons meaning that hydrogen can have isotopes. Where hydrogen has one proton but no neutrons in its nucleus it is the isotope hydrogen-1 (H-1). The numbering of the isotope refers to the total number of protons and neutrons. If it has one proton and a single neutron, it is isotope hydrogen-2 (H-2) and is called deuterium. If it has one proton and two neutrons it is isotope hydrogen-3 (H-3) and is called tritium. As another example, the element carbon, which is the basis of all life on earth, has six protons and six neutrons in its nucleus, hence atomic number 6 and has six electrons in orbit to balance the proton charge. Carbon can exist as any one of three isotopes. The most common form of carbon has six protons and six neutrons in which case it is the isotope carbon-12 (C-12). If it has seven neutrons it is the isotope carbon-13 (C-13) or eight neutrons isotope carbon-14 (C-14). In 1933, the heaviest known naturally-occurring element was uranium having atomic number 92 which existed as isotopes uranium-235 having 92 protons and 143 neutrons (U-235), uranium-236 with 144 neutrons (U-236), uranium 237 with 145 neutrons (U-237) and uranium 238 with 146 neutrons (U-238).

The number of neutrons can affect the stability of an isotope. C-12 is stable but C-14 undergoes slow decay at a rate called its half-life which is time it takes for half of it to decay. So after 5,730 years half of the C-14 will have decayed and after another 5,730 years half of the remainder will have decayed leaving a quarter of the original amount and after another 5,730 years only an eighth will remain. The decay process involves one of the neutrons in the C-14 nucleus emitting an electron and turning into a proton so instead of six protons and eight neutrons it now has seven protons and seven neutrons making nitrogen-14 (N-14) with atomic number 7 which is a completely different element. The black, sooty powder of carbon has turned into a colourless, inert gas.

Because of their positive electrical charge, protons normally repel each other but where elements are forged at the centre of stars the massive gravitational forces press them so close together that they become trapped by a force that is strong enough to overcome the repelling electromagnetic force trying to drive them apart. This is called the 'strong force' and is the strongest force in nature, but its effect is only felt at the level of atomic nuclei and it acts only over extremely short distances (0.000000000000001m). It is this strong force that holds all of the components of an atomic nucleus together. In a stable isotope, C-12 for instance, the strong force is sufficient to prevent any sort of atomic decay under normal circumstances but in some isotopes such as C-14 the components of the nucleus, which are in constant motion, become unbalanced and decay occurs with the emission of radioactivity. Radioactivity can be dangerous for life forms depending on the energy levels of the particles emitted. Many natural substances, such as granite, are known to emit low levels of radiation but very high levels can be produced in laboratory experiments by forcibly 'splitting' atomic nuclei and it was this work that Enrico Fermi was conducting in the early 1930s.

Chapter 2

SPLITTING THE ATOM

'Every scrap of solid matter in the world would become an available reservoir of concentrated force.'

H.G. Wells, *The World Set Free*

The story of the atomic bomb really starts with the research of Jean Frédéric Joliot and his wife Irène in Paris and that of Enrico Fermi in Rome. Frédéric had started work at the Marie Curie Institute in Paris in 1924 when he met and married another scientist, Irène, the daughter of the pioneering physicists Marie and Pierre Curie. Irène had worked briefly as a nurse alongside her mother on the First World War battlefields and after the war continued her studies in mathematics and physics at the Sorbonne. 'We had very different personalities,' Irène said later of Frédéric, 'but . . . in work and in life . . . we complemented each other.'[1] They would build upon work done by Irène's parents as far back as 1896 on the radioactive properties of uranium and polonium but at the time it was radium that was of greatest interest. The Curies devoted the rest of their lives to further experimentation with that element and it was the intense radioactivity of radium that would eventually have such a deleterious effect on their health. Frédéric and Irène would later both change their names to Joliot-Curie but continued to sign their scientific papers Irène Curie and Frédéric Joliot.

In the course of their research, the Joliot-Curies bombarded certain elements such as aluminium and magnesium with alpha particles (an alpha particle is the nucleus of a helium-4 [He-4] atom made up of two protons and two neutrons) and they obtained unstable radioactive isotopes. They had begun working collaboratively in 1929 and published a paper in 1932 in which they claimed to have irradiated paraffin wax with alpha particles and observed, to their astonishment, an output of radiation three times more powerful than the radiation put in. This was so far outside current scientific thought that many leading scientists of the day, such as Niels Bohr, Enrico Fermi, Werner Heisenberg and Lise Meitner, challenged the results. Back in their laboratory, determined to verify their earlier findings, they instead made a startling discovery. A thin sheet of aluminium had been bombarded with alpha particles but when the bombardment ceased, after a short delay, the aluminium started emitting a new kind of radiation that left traces inside an apparatus known as a cloud chamber. This seemed to connect with work done back in 1917 by Rutherford when he had demonstrated that it was possible to convert the element nitrogen into oxygen (atomic number 8) by bombarding it with alpha particles. In this case the nitrogen nucleus had absorbed an alpha particle but the Joliot-Curies had gone one step further by bombarding aluminium (atomic number 13) with alpha particles and transmuted it into a different element, in this case phosphorus (atomic number 15) which proved to be inherently unstable and slowly released radiation.

For this work, the couple were to win the Nobel Prize for Chemistry in 1935 for 'Recognition of their synthesis of new radioactive elements'. Only one other married couple had been jointly awarded a Nobel Prize before and that had been Irène's parents. In his Nobel acceptance speech, Frédéric warned that artificially produced radioactivity might one day lead to 'transmutations of an explosive character [in a] chain reaction'.[2] A nuclear chain reaction occurs when one single nuclear reaction causes an average of one or more subsequent nuclear reactions, thus leading to the possibility of a self-propagating and rapidly expanding series of these reactions. In 1937, Frédéric was appointed Professor of Nuclear Chemistry at the Collège de France where he supervised the installation of a seven million electron volt Van de Graaff cyclotron (particle accelerator) while Irène, now in increasingly poor health due to tuberculosis, devoted her time largely to the upbringing of their children, Hélène and Pierre, after having briefly been appointed to the post of Undersecretary of State for Scientific Research in the French Popular Front government of 1936. In January 1939, Frédéric went on to establish a research programme at the Radium Institute in Paris. It was at this time also that the Joliot-Curies reacted to the threat from Nazi Germany by ceasing publication of their research and sealing all their results and secreting them in the vaults of the Académie des Sciences where they remained concealed until the end of the Second World War.

Following on from the discovery of artificial radioactivity by the Joliot-Curies, Enrico Fermi, a professor of theoretical physics at the

Sapienza University of Rome, used neutrons in his experiments instead of alpha particles. The problem with the Joliot-Curie method was that both the particles used for bombardment (protons and alpha particles) and the nucleus itself were positively charged. This meant that there was a powerful repulsion between the similarly-charged projectiles and the target, making it extremely difficult for them to come together. In fact the closer they came together the stronger the repulsion, so in order to overcome the electrical barrier of the nucleus the protons or alpha particles had to be accelerated to very high speeds in a cyclotron. The neutron, however, with no electrical charge, could penetrate a nucleus at much lower speeds and with much less expenditure of energy, making it a much more practical tool for nuclear exploration. When a neutron struck a nucleus it was 'about as catastrophic as the moon [striking] the earth' and the nucleus was violently shaken up as a result.[3]

Fermi was able to demonstrate that nuclear transformation occurs in almost every element subjected to neutron bombardment and he even managed to split the uranium nucleus, although at the time the way he had set up his experiment meant that he did not realise what he had done. Unable to separate out the product of this bombardment, he could not test its properties and he assumed that rather than splitting the uranium nucleus, which from a theoretical point of view at the time was considered most unlikely, he had created a new artificial 'transuranic' element with atomic number 93. Fermi had also, surprisingly, found that the transformations were speeded up if the bombarding neutrons were slowed down by first passing them through a substance such as paraffin wax. As was the case generally with science, these results were questioned, especially when there was little corroborating proof. One scientist, the German chemist Ida Noddack, actually suggested that the uranium had been shattered into two or more much smaller fragments and it was these that Fermi had detected but she was simply ignored because existing nuclear theory did not allow it and Noddack had failed to offer experimental proof or a theoretical basis for this possibility. Her ideas were mocked as being 'ridiculous' by such as Otto Hahn who, as we shall see, was to conduct the very experiments that would later prove her to have been right. It cannot be discounted that Noddack's theory was not taken seriously and was not tested at the time simply because she lacked scientific standing because of her gender.

On 10 November 1938, Fermi would be awarded the Nobel Prize in Physics for his 'Identification of new radioactive elements and his discovery, made in connection with this work, of nuclear reactions effected by slow neutrons'. He had married Laura Capon in 1928 and they had two children but when Mussolini brought in racial laws aping the Nazis, Laura's Jewish origins put them all in danger. After travelling to Stockholm for the Nobel ceremony, Fermi and his family did not return to Italy but chose instead to emigrate to the USA. They would arrive in New York on 2 January 1939 when Fermi was immediately offered positions at five universities and accepted one at Columbia University in New York where he had already given summer lectures in 1936.

The Kaiser Wilhelm Institute (KWI) in Berlin was at the forefront of scientific research during the 1930s and employed some of the world's greatest atomic scientists, one of whom was the Viennese Lise Meitner. At the age of 14, Elise (Lise) Meitner had been a proficient but 'not expert' pianist who could not compete with her gifted sister Auguste but she showed considerable aptitude in other fields than music and was already keeping copious records of her 'scientific experiments'. Denied by law from any further state education because of her gender, Lise was forced to leave school but for the next five years studied at home and dreamed of going to university, which at the time was unlikely since university education was denied to girls in Austria. Fortunately, the law was changed and her lawyer father Philip paid for another two years of private tuition so that she could prepare for the entrance examination.[4] She passed and began studying physics at the University of Vienna in 1901 and arrived in Berlin with her doctorate in 1907 to continue her studies but was shocked to discover that in Prussia women were still not admitted to its universities. She was, however, allowed to unofficially attend lectures given by Max Planck and it was here that she first met her lifetime collaborator Otto Hahn, who gave her work in his chemistry laboratories within the University of Berlin, albeit as a guest without remuneration. However, on the somewhat hilarious assumption that her long hair made her a 'fire hazard' when working on her experiments, she was forced to work in a converted carpenter's shop in the basement and forbidden to enter the main university buildings at all. When the eminent British scientist Sir Ernest Rutherford visited Hahn's laboratories, Meitner was obliged to take Rutherford's wife on a Christmas shopping trip while the two men discussed the latest developments in physics.[5]

Things improved somewhat in 1912 when Hahn and Meitner moved to the new Kaiser Wilhelm Institute for Chemistry (KWIC) to study the exciting field of radioactivity in pursuit of knowledge about the internal structure of atoms. Meitner was officially designated as an assistant to Planck which eventually meant that she now had an official paid position. After serving on the Austrian front as an X-ray technician during the First World War, Meitner returned to work with Hahn in 1917 and together they won the Liebnitz Medal of the Berlin Academy of Sciences for their discovery of the radioactive decay of protactinium. When their paper 'The Mother Substance of Actinium, a New Radioactive Element of Long Half-life' was published, although Lise had done most of the work it was Hahn's name that appeared first. Her intellect and inspired theoretical reasoning, however, soon earned her the respect of her peers, so much so that by 1926 the 'rather shy and withdrawn' Meitner had become the first female professor of Physics in Germany and by 1932 was generally considered to be one of the country's foremost nuclear scientists.

But then everything changed when Adolf Hitler was appointed Chancellor of Germany in January 1933. The Law for the Restoration of the Professional Civil Service was passed excluding Jews and other non-Aryans from various professions and organisations, including from having roles in universities. The position of the Jewish Meitner came under threat but her Austrian nationality, for the moment, was some protection from immediate persecution. However, her presence at the KWI was the cause of some concern for Hahn who told Meitner that it might be better if she left the Institute before they were all marked out for attention by the Gestapo. Bravely, Meitner resisted. She saw no reason for such a drastic step that would separate her from the work to which she had devoted her life. She refused to be driven out and continued to turn up for work. Hahn was torn between his sympathy for her and what he saw as his duty to the KWI which might become a target for the notoriously anti-intellectual Nazis. At the same time, Hahn was discussing with colleagues ways of organising mass protests by scientists and intellectuals to prevent the catastrophe that was facing German science but the general mood was that things had already gone too far and if leading scientists resigned from their posts many younger, eager and politicised ones would be only too quick to take their places.

Hahn naively hoped that the hounding of the Jews might abate and managed to persuade the KWI directors to keep Meitner on the staff but he was unable to avoid sanctions completely. She was dismissed from the faculty of the University of Berlin and barred from speaking at scientific meetings. All they could do was press on and hope that the political situation would calm down, which at the time actually showed no signs of doing so. Trying to ignore the danger beyond their laboratory, Hahn the experimenter, at this time 'unquestionably the ablest radiochemist in the world',[6] and Meitner, the theorist, would go on to build on the foundation of work done by Fermi in Rome and that of the Joliot-Curies in Paris. Assisted by a young German chemist, Fritz Strassmann, they began looking for evidence for new types of radioactive substances by studying the effects of the slow neutron irradiation of uranium.

For five years, Meitner was able to maintain a low profile and avoid trouble but after the *Anschluss,* Austrian Jews came much more under the malevolent glare of the Gestapo and the SS. Professor Kurt Hess at the KWI made it quite clear where his sympathies lay when he called Meitner 'the Jewess who endangers the Institute'.[7] Offers came in from abroad for her to emigrate, including one from Niels Bohr, the world-renowned director of the Institute for Theoretical Physics in Copenhagen. The idea was that she would be invited to attend a week-long conference there and would simply stay behind when all the others left. It was a tempting offer but Meitner refused to take advantage of it, hoping against hope that somehow she would be allowed to stay in Berlin and continue working with Hahn. The question arises as to the extent to which she was so engrossed in her work that she refused to admit what was happening in Germany or whether she deluded herself into thinking that her Austrian passport would protect her and allow her to operate 'under the radar' and stay with Hahn, with whom she had a close and productive working relationship.

It was not long before reality kicked in though. Events caught up with her when she was denounced to the Gestapo by her neighbour for being Jewish and once again she came under the spotlight. The offer from Bohr now seemed to be the only way out but when she went to the Danish embassy to get a visa they turned her away because the *Anschluss* had rendered her Austrian passport invalid. Bohr came to Berlin to see what he could do to help but he left empty-handed. Dirk Coster, the professor of Physics at Groningen University in the Netherlands, invited her to join him there. Local Dutch scientists had offered to club together to pay her a modest stipend but on 16 June, Carl Bosch, the Director of KWI, received a letter from the German authorities saying specifically that Heinrich Himmler himself found it 'undesirable that well-known Jews [are allowed to] leave Germany to travel abroad where they appear to be representatives of German science or with their names and their corresponding experience might even demonstrate their inner attitude against Germany'. Meitner was clearly not going to be allowed to leave and while she may have been able to continue with her work for the present, in the long term her fate would be sealed.[8]

Bohr continued to lobby Danish and Dutch scientific institutions to offer Meitner a place but the only one that looked possible was the newly-formed Nobel Institute Manne Siegbahn in Stockholm, which could only offer 'small' possibilities. After weighing up her options, Meitner accepted the offer from Sweden but on 4 July word came that all German borders were to be closed after which all chances of her leaving legally would disappear. Still there was no confirmation that Sweden would let her in even if she could escape Germany, but Coster had managed to get authorisation for her to enter the Netherlands. On 11 July, he came to Berlin. The next day Meitner went to work as usual but stayed late and, in the evening, went home and with Hahn's help packed a few personal belongings ready to go with Coster back to Groningen. Meitner had been informed by the Dutch Ministry of Justice and the office of the Dutch Border Guards that 'provisional admission' would be granted but that still left the problem of the German border police. Although written confirmation of this would be provided, Hahn admitted to 'shaking with fear' that Meitner would be turned back at the border and probably arrested.[9] He slipped her a small diamond ring that had belonged to his mother so that she might have something to trade in an emergency.

Hahn did not have a car so the British spy and editor of the German scientific journal *Die Naturwissenschaften* Paul Rosbaud drove her to the station. Rosbaud's Jewish wife and daughter had fled to Britain, but he remained in Germany during the war and kept in close touch with scientists there. In this way he was able to pass information to Britain about German nuclear research and later about development of the V2 rocket. Rosbaud was also able to compile a list of German scientists who 'kept their moral integrity all during the Nazi regime and who restricted themselves to basic research'. Interestingly, in 1942 Heisenberg would warn his wife, Elizabeth, to be discrete in Rosbaud's presence. 'He is a spy', he told her, 'but I don't know for which side.'[10]

On the train Meitner met up with Coster in what they hoped looked like a chance encounter. At the Dutch border town of Nieuwe Schans, a Nazi military patrol halted the train and passed through the carriages checking the passengers' documentation. All Meitner had to show was her invalidated Austrian passport. The soldiers looked at the 'petite lady, alighting from the train, trembling with fright', took her passport and disappeared.[11] Her whole life hinged on that moment. She later recalled that '[she] got so frightened, [her] heart almost stopped beating'. For ten minutes, which seemed like 'so many hours', she sat and waited then a guard returned, handed back her passport 'without a word' and let her get back on the train. On their arrival in Groningen,

Coster sent a coded telegram back to Hahn, to say that the 'baby' had arrived. On 26 July, having got a Swedish visa at last, Meitner flew to Copenhagen where she spent time with Bohr and her nephew Otto Robert Frisch before going on by train to Kungälv in Sweden.[12] When questioned, Hahn told the authorities that Meitner had gone to visit her family in Vienna.

In Sweden, in the almost deserted Nobel Institute for Experimental Physics, Meitner was left to fend for herself with hardly any equipment, no assistant and almost no communication with anyone. Meitner's director, Manne Siegbahn, pleading poverty, paid her a measly salary and treated her 'as if she were invisible' but still she continued her work, which was essentially theoretical, as best she could.[13] All in all, in Stockholm she found herself isolated 'like a wind-up doll running on automatic [and] empty of real life'.[14] Members of Siegbahn's group, having no understanding of the displacement and anxiety common to all refugees, did little to make her feel comfortable and generally looked upon her as an outsider. She became withdrawn and depressed from the trauma of losing friends and relatives to persecution back home in Germany and Austria and felt the exceptional isolation of a woman who had single-mindedly devoted her life to her work. She tried with limited success to get her personal effects sent on from Berlin but she was unable to transfer any money from her German bank account. She had lost almost everything and was starting her life over again in the direst circumstances one year short of her 60th birthday.

Hahn, Meitner and Strassmann had irradiated uranium with slow neutrons back in 1935, hoping to create new elements beyond uranium which they called eka-rhenium (beyond rhenium) and eka-osmium. While Hahn and Strassmann had done the practical laboratory work, Meitner had tried to work out what they had achieved from a theoretical standpoint but the result defied rational explanation. Unable to progress on that front, they turned to other work but their interest in it was rekindled in 1937 when the Joliot-Curies in Paris claimed to have produced thorium through irradiation of uranium. Meitner wrote to them to dispute their findings from a theoretical perspective but further work in their own laboratory began to convince the German team that the French had been right all along. After Meitner had gone to Sweden, Hahn and Strassmann continued their experiments but in the absence of their theorist were now even more stumped when it came to finding an explanation of their results.

On 19 December 1938 Hahn wrote to Meitner describing how he and Strassmann had bombarded the nucleus of a uranium atom with a neutron and discovered something 'so strange' going on when it appeared to transmute into much lighter elements rather than nearby elements such as radium, which is what would be expected according to current theory. These lighter products had the properties of barium and lanthanum instead and suggested that the uranium atom had been split into two smaller atoms. There was nothing in the prevailing theory to explain how such a thing might happen so he was lost for an understanding of what he had observed in the laboratory. Believing that the uranium simply could not break up, he frustratedly asked Meitner, 'perhaps you can suggest some fantastic explanation' and signed off the letter with his 'very warm greetings'. Hahn knew that if he was right then he had opened up a whole new realm of physics that might have revolutionary consequences but, having castigated the Joliot-Curies back in 1937 for what he had considered to be sloppy work and unreliable results, he needed his theorist, Meitner, to take a look and make sure that it was not he who would be subjected to obloquy this time. Meitner read the letter and wrote back immediately saying that the results were 'very difficult to accept' but the understanding of nuclear physics was progressing at such a pace that nothing could be ruled out as 'impossible', signing off with 'much love and the very best for the New Year'.[15] The two very obviously continued to enjoy a warm and friendly relationship. Hahn confided to Meitner in his next letter that his results were 'absurd' but he felt obliged to publish them, hoping that she could come up with an explanation.

Meitner had arranged to spend Christmas with her nephew, the Austrian émigré physicist Otto Frisch, at a hotel in Göteborg. Like his aunt, Frisch had been born in Vienna. He had studied in the Universities of Berlin and Hamburg before fleeing from Nazi racial laws in 1933 when he went to the Institute for Theoretical Physics in Copenhagen but he did not stay there long before moving on to Birkbeck College in London. He then went back to Copenhagen where he worked with Niels Bohr.

Meitner and Frisch took long walks in the snowy Swedish countryside and applied all their knowledge and inspiration to trying to explain what must have happened for Hahn to get such a result. Frisch found it hard to believe that Hahn had overcome the powerful nuclear forces that held the nucleus together and initially called his conclusion 'a mistake' but Meitner had every confidence in Hahn's skill and efficiency as an experimenter.

The question was how the impact of a single neutron had caused the uranium nucleus, which was massive by comparison, to be sundered. It was like saying that one could split an apple apart by firing a pea at it. The uranium nucleus, however, was not a solid object like an apple: it was made up of many small sub-atomic particles. Furthermore, because the uranium nucleus was so massive it was inherently unstable as the electrical forces of its many constituents, relentlessly trying to break it apart, vied for dominance over the strong nuclear force that held it all together. Maybe the addition of a single neutron was enough to tip the balance and disturb this uneasy truce between the forces sufficiently to cause a ripple of extra energy that might just agitate the nucleus enough to drive it over the edge of instability, causing it to split asunder. An explanation of just how this could happen required Meitner and Frisch to dredge up all their accumulated knowledge and apply their formidable intellects. They were forced to go beyond conventional theory into the realm of speculation and eventually came up with the idea that the extra neutron absorbed by the uranium nucleus might have exaggerated pre-existing tidal fluctuations within the nucleus. They applied the Bohr model of an atomic nucleus which saw it acting not as a solid object but more as a bubble that could be distorted from within to stretch into something like a dumbbell shape.

Assuming this to be the case, disturbances caused by the extra neutron could possibly, according to their calculations, be just enough to pull the protons in the nucleus far enough apart so that their natural electrical repulsion would overcome the immensely powerful but short-range strong nuclear force that normally held it together. If this was to occur, said Meitner, then two separate particles of similar charge would result and they would repel each other and fly apart with great energy, but it was still a mystery what the source of that energy would be. Meitner then recalled a lecture she had attended in 1909 in Salzburg delivered by Albert Einstein during which he had used his Theory of Relativity to derive the famous equation $E = mc^2$, demonstrating the relationship between energy and matter, and it was this that proved to be the key to unlocking the mystery. Einstein had shown that mass and energy were interchangeable, where one could become the other, and a small amount of mass, if converted completely, would release a phenomenal amount of energy. Was it possible that the splitting of the uranium nucleus had resulted in a very small amount of mass converting into pure kinetic energy? Frisch likened the process that he and Meitner had postulated to biological cell division and used the same term 'fission' to describe it.[16] Fission now became the accepted term to describe the process by which a small number of neutrons are given enough energy to collide with and split an atomic nucleus.

Eager for Meitner's 'frank opinion', Hahn wrote again concerning his 'barium fantasies', asking whether it was possible that a neutron could change into a proton by casting off an electron. Meitner was willing only to cautiously reply that she considered it no more than a possibility that his results would stand up to scrutiny given that they were 'against all previous laws of physics'. She did not want Hahn to suffer the indignity of having his experimental skills questioned. Hahn refined and repeated his experiments but came up with the same results. He wrote again to Meitner with the new set of results. Her total confidence in his soundness as an experimenter gave her no option but to accept his findings and she replied praising his 'wonderful result' and the 'beautiful field of work ahead of you' but her chagrin at being debarred from sharing it had left her 'with very empty hands'.[17]

Hahn and Strassmann published their findings in *Die Naturwissenschaften* in January 1939 but omitted any mention of Meitner's crucial contribution. Several reasons have been put forward to explain why Hahn did this and why he did so little to protect her from Nazi persecution resulting in her exile, but all are speculative and Hahn himself gave no explanation. Meitner and Frisch published their own report to the wider scientific community in *Nature* on 11 February in what was hailed as a brilliant extension of existing nuclear theory.[18]

Frisch now asked himself that if U-235 was naturally radioactive and was continuously undergoing decay why had all the uranium in the earth not decayed or been destroyed by explosion long ago. If it was decaying and emitting neutrons during its decay, why did those neutrons not go on to strike other U-235 nuclei and cause them to decay and so on until it was all gone? He quickly saw that because uranium was deposited in ores mixed with other minerals, most of U-235 nuclei would be shielded from direct strikes so there was no possibility of a chain reaction. With that, he recalled the moment when an 'exciting vision arose that by assembling enough pure uranium (with appropriate care!) one might start a controlled chain reaction and liberate nuclear energy for peaceful purposes on a scale previously unheard of. Of course, the spectre of the bomb [lurked in the background].'[19]

In February 1939, Meitner wrote to Hahn to express her feelings at not being included as co-author of Hahn's *Die Naturwissenschaften* paper. She doubted that her current employer Siegbahn would ever come to appreciate her as a valuable asset in his research centre and he would now be convinced that 'I never did anything and that you also did all the physics in Dahlem [KWI]'. Her misery and despondency were evident when she wrote 'I am gradually losing all my courage. Forgive this unhappy letter. I never wrote how bad it really is.' To her brother Walter she wrote 'much as these results make me happy for Hahn, both personally and scientifically, many people here must think

I contributed absolutely nothing to it—and now I am so discouraged; although I believe I used to do good work, now I have lost my self-confidence.'[20] Hahn was awarded the Nobel Prize in Chemistry in 1944 for the discovery of the fission of heavy nuclei but he did not receive it until 1945. The award was and still remains controversial, primarily because Hahn's Berlin colleagues, Strassmann and Meitner, were not included in the citation.

Later, on 9 April 1940 when the Germans invaded Denmark, Meitner had only just arrived on a visit to Bohr's laboratory in Copenhagen, which was ransacked by German troops. She returned to Sweden where her miserable isolated existence continued with 'the great difficulties in the laboratory and Siegbahn's lack of interest in her scientific work'. She would write to say that she 'longs for a meadow full with spring flowers in the Alps'.[21] She would later receive a number of invitations to emigrate to the USA and work at Los Alamos after the science had been put at the beck and call of the military, but she angrily replied that she 'would have nothing to do with a bomb'. She eventually arrived in New York on 25 January 1946. After the war she wrote to Hahn in a spirit of 'true friendship' but, in the letter, was very critical of what she saw as his betrayal of his friends and his country under the Nazis. Despite this, Meitner retained a 'deep affection' for Hahn and the two remained lifelong friends. Much of her time and energy in the immediate post-war period was spent trying to persuade her former colleagues in Germany to acknowledge and accept their failure of responsibility by staying quiet during the horrors that had taken place in the Third Reich.

It took Hahn many years to admit his own culpability. In 1958, he wrote to Meitner on her 80th birthday: 'We all knew that injustice was taking place, but we didn't want to see it, we deceived ourselves . . . Come the year 1933 I followed a flag that we should have torn down immediately. I did not do so and now I must bear responsibility for it.' He thanked Meitner 'for trying to make us understand, for guiding us with remarkable tact' towards a recognition of the way the German physicists had worked under, and in many ways for, a monstrous regime. Meitner was nominated twenty-nine times for a Nobel Prize but was never awarded it. Her nomination in 1945 was actually blocked by Manne Siegbahn which is a graphic example of the misery she must have experienced working under him. She died on 27 October 1968 in Cambridge, England and lies buried in the village of Bramley, Hampshire. The inscription on her gravestone reads, 'Lise Meitner: a physicist who never lost her humanity'. A residential street in the village, Meitner Close, is named after her.

After his meeting with Meitner at Kungälv over Christmas in 1938, Frisch had returned to Copenhagen where, given the enormity of what had been discussed, he somewhat hesitantly apprised Bohr of Hahn's observations and the theoretical explanation that he and Meitner had proposed. If he had expected Bohr to dismiss their ideas, he was relieved and surprised when Bohr 'struck his forehead with his hand and cried out "Oh what idiots we have all been"'.[22] Frisch wrote to Meitner that Bohr had been 'in every respect in agreement with us'. Frisch then conducted a frantic telephone correspondence with Meitner to flesh out their theory and had only partly finished a paper by the time Bohr was ready to leave for the USA where he was taking up a temporary post at the Institute for Advanced Study at Princeton. Bohr agreed to refrain from discussing the idea with anyone until Frisch and Meitner's paper was published which he expected to happen while he was in mid-Atlantic on the SS *Drottningholm* but they must have known how difficult that would be for the garrulous Bohr.

Whilst Bohr had every intention of withholding the news of the discovery until both Hahn and Meitner had published, his great enthusiasm for the new ideas had caused him to inadvertently let it slip by telling one other person on the voyage over and failing to swear them to secrecy. The news burst out and spread like wildfire once the *Drottningholm* docked but not all who heard it were ready to accept it at face value. The physicist George Gamov, for instance, thought that Bohr had 'gone crazy'. When Bohr got up to speak at a conference at George Washington University on 26 January the audience already knew what he was going to say but remained surprisingly subdued. In the few days that the US scientists had been digesting the new ideas it had become immediately obvious that these theoretical findings had implications in the field of weapons research and had created a great deal of speculation over the possibility of Germany developing the techniques required to weaponise the science and divert the new findings into a cogent strategy to build an atomic bomb of fearsome power. The results of Bohr's official revelation were immense. By now the simple logic of what he was saying was undeniable. A flurry of activity broke out when he had finished speaking and many scientists returned to their laboratories and immediately set about replicating Hahn's experiments and confirming his results. It dawned on many that Fermi could have revealed the same startling results as far back as 1933 had he set up his experiments slightly differently. Because he was such a perfectionist, Fermi had covered samples of uranium with a thin sheet of inert material to prevent the escape of particles which his theories had convinced him were unimportant, but it was just these particles escaping from his sample that would have opened his eyes to a whole new concept of nuclear physics had he detected and analysed them. Had he not allowed himself to make these false assumptions, the crucial breakthrough that Hahn and Strassmann made in 1938 might have been made early enough for both Germany and the Soviet Union to develop nuclear technology prior to the outbreak of the Second World War which would have had profound consequences.

Bohr went on to postulate that the isotope U-235 was the most suitable material for fission to create a bomb with a much less stable nuclear composition than the U-238 isotope that made up much of the naturally-occurring mineral ores but, given the 'extreme difficulty' in separating U-235 from U-238 since they had similar chemical properties, he considered the whole idea of translating his 'theoretical possibility' into a reality was beyond 'the entire effort of a country'.[23]

The nucleus of an element is said to have a fission barrier which is a measure of the energy a neutron must possess if it is to force its way close enough to the nucleus for the powerful short force to ensnare it and bind it. As Hahn showed, nuclear fission can occur when a nucleus is rendered unstable by the capture of a passing neutron. The probability for a fission reaction taking place, however, depends on how much kinetic energy (energy of motion) there is in the neutrons which may be classified as fast or slow.

If slow neutrons are fired at a target of natural uranium oxide (99.3 per cent U-238 and 0.7 per cent U-235) they will be absorbed by the U-235 because each of its atoms has an odd number of neutrons (143) and, according to the atomic exclusion principle, has a low threshold over which a neutron must pass to be absorbed. U-238, on the other hand, with an even number of neutrons (146), will only allow fast neutrons to cross its threshold. If the mixture continues to be bombarded by slow neutrons, they will all end up being absorbed by the U-235 and the U-238 will be unaffected. When a nucleus of U-235 absorbs a slow neutron, however, it undergoes fission, breaks up into two smaller elements and releases two or three fast (thermal) neutrons which will now be preferentially absorbed by the vastly more abundant U-238 which does not undergo fission. The U-235 will therefore become starved of slow neutrons and cannot give rise to a chain reaction. To overcome this restriction, the neutrons emitted through U-235 fission must be slowed down so that the U-238 will not absorb them. Then all the neutrons produced by fission of U-235 will be available for absorption by other U-235 nuclei and an explosive chain reaction will result if there is a critical mass of U-235. A critical mass is the minimum amount of fissile material required to ensure that there are enough collisions between nuclei of the material and free neutrons to start a chain reaction. However, having a quantity of fissile material in excess of its critical mass would not necessarily cause an explosive chain reaction to start. For that to happen, the critical mass must be arrived at almost instantaneously, as would occur if two sub-critical masses were fired towards each other to collide at high velocity.

As stated, in naturally-occurring uranium oxide, commonly referred to as yellowcake, when U-235 fissions it emits fast (thermal) neutrons, which are captured by the U-238 but in February 1939, Bohr published an article in the peer-reviewed scientific journal *Physical Review* showing that if the neutrons were slowed down by using a 'moderator' then a self-sustaining chain reaction might be controlled. Now for a chain reaction to be self-sustaining, at least one neutron emitted by the U-235 has to penetrate another U-235 atom, the prospect of which is increased as the purity of the U-235 goes up. Many neutrons will escape altogether into the surroundings and others may be absorbed by impurities within the fissile material. This is why uranium does not fission in its natural state; the U-235 is not sufficiently concentrated. It would be the physical separating out of U-235 from yellowcake that would prove to be one of the most intractable problems scientists faced. There would follow the question of exactly what concentration the U-235 would have to be raised to and how much of it would be required to achieve a critical mass. It was held for some time in the early stages of atomic bomb development that a critical mass of uranium would need to be quite large, possibly in the region of tons, which was far more than had ever been produced. A twofold approach was envisaged whereby on the one hand the methods of purifying U-235 would be examined and on the other whether or not it was possible to bring about a chain reaction using unprocessed yellowcake. Neither gave much prospect of early resolution.

To maintain and sustain a controlled nuclear reaction, each neutron-uranium nuclear collision must result in just one neutron being released for interaction with another uranium nucleus. If this ratio is less than one, then the reaction will fizzle out; if it is greater than one the rate of reaction will grow uncontrolled and at extremely high speed, resulting in an atomic explosion. It was shown experimentally that each neutron-uranium collision produces on average between two and three thermal neutrons so if the reaction is to be controlled some means of moderating the availability of these neutrons must be found by introducing a neutron- absorbing material. This would allow scientists to start, run or stop the reaction as they chose.

Another issue with neutrons, as seen earlier, is that they often have too much kinetic energy and so must be slowed down to ensure that they are absorbed by U-235 and this is done through the use of a moderator such as heavy water or graphite. The probability that fission will occur is determined by the neutron cross-section for that reaction which, in simple terms, is the area surrounding the target nucleus within which the incoming neutron must pass if the reaction is to take place and this varies according to the speed of the neutron. Once the fast neutrons have been slowed, they are more likely to produce further nuclear fissions or be absorbed by the control rod. It was Bohr and John Wheeler working at Princeton who would introduce the idea of a moderator to slow down the neutrons so that their energy levels would be too low to break into the U-238 nuclei and so more of them would be available for absorption by the U-235. Only very few materials with a very specific set of properties proved to be efficient moderators. They cannot absorb neutrons themselves but the closer in size the nucleus of an atom is to a neutron, the more the neutron will be slowed and the more effective it will be as a moderator. For this reason, lighter elements tend to be more efficient moderators. One suitable candidate as moderator was heavy water. A molecule of normal water contains two hydrogen and one oxygen atom (H_2O) whereas heavy water (D_2O) contains two deuterium atoms and one oxygen atom. You will recall that deuterium differs from hydrogen in that whereas a hydrogen atom contains one proton and one electron, deuterium contains one proton, one electron and an extra neutron. Heavy water exists naturally in very small quantities in normal water and is separated out by a complex industrial process whereby naturally-occurring water is electrolysed to remove most of the hydrogen, a process which requires very large supplies of electricity, making heavy water a very expensive commodity.

In February 1939, after reading Hahn and Strassmann's article in *Die Naturwissenschaften*, Fermi and Leo Szilard collaborated with Walter H. Zinn and Herbert L. Anderson at Columbia to investigating the conditions required to start a nuclear chain reaction in each of the three isotopes of uranium. Fermi had guessed, and hoped, that it would be the abundant heavier U-238 that fissioned most freely, but it turned out to be the rare U-235 as Bohr had predicted in his 15 March 1939 article in *Physical Review*. Their results confirmed that U-235 was fissionable by both slow and fast neutrons, although it was achieved with more control by the former. A chain reaction might be possible because additional neutrons were emitted when the uranium underwent fission which, of itself, was significant because it confirmed that nuclear power was a reality but did not necessarily indicate that it would lead to development of an atomic bomb. In fact, Fermi said that there was 'little proof that we were not pursuing a chimera'. However, the Danish scientist and assistant to Niels Bohr, Christian Møller, confirmed that the lighter elements produced by the fission of uranium would not have sufficient energy to hold onto all the neutrons that had been present in the original uranium nucleus and therefore a number of free neutrons would also be released in the process. It was Møller's view that these free neutrons would indeed have sufficient energy to collide with further atoms of uranium, if present, and repeat the fission process in a chain reaction which would be the precursor to an explosion.

Leo Szilard was born in Budapest and had worked alongside Einstein in the Institute of Technology in Berlin where they had jointly patented an electromagnetic refrigeration pump that was sold to AEG but soon abandoned as unworkable.[24] He had conceived the idea of a nuclear chain reaction as early as 1933 in a 'visionary conjecture'[25] but received little support when he moved to England and shared his idea with Rutherford who dismissed it out of hand. Not only had Szilard come up with the idea of a chain reaction but he had also considered what critical mass of fissionable material would be required for the chain reaction to be self-sustaining. He firmly believed that if he could find a suitable fissile material it was only a matter of time before he could produce an explosion. He later applied for a patent for a reactor device that could be adapted to make a bomb and handed it over to the British Admiralty to be securely archived but he doubted that such a bomb could be constructed without a huge amount of fissile material, making it so massive that it could not possibly be carried by any aircraft in production at that time. He was convinced, however, that should such a bomb be produced it would 'go far beyond all military conceptions' and he would later say so in a memorandum attached to Einstein's famous letter to Roosevelt of 15 August 1939.[26] Having made several visits to the US over the years when he collaborated with Fermi, Szilard made a permanent move to New York and took US citizenship in 1940. Working full-time with Fermi at Columbia, he published results of his own experiments showing that the rate of neutron production during uranium fission was indeed sufficient to generate a chain reaction.

At the same time the Austrian-born Hans von Halban and Frédéric Joliot-Curie, together with Francis Perrin and Bruno Pontecorvo and the Russian-born Lew Kowarski, published similar findings to those of Szilard and Fermi in *Nature* on 19 March 1939 postulating that fission would produce new neutrons that could result in a chain reaction. These scientists now working in France concentrated on calculating the critical mass of uranium required to start and sustain such a chain reaction. They too concluded that a bomb would require many tons of fissile material and applied for three patents on behalf of the Caisse Nationale de la Recherche Scientifique at the beginning of May 1939. The first concerned the technical principles of nuclear energy production, the second involved absorption of neutrons in the stabilisation of a chain reaction and the third addressed the principles of critical mass and nuclear explosions. At the time, however, the team were focused on slow neutrons in a reactor by introducing a moderator and it was here that the role of heavy water and graphite became so important to them.

The French Minister of Armaments, Raoul Dautry, gave his full support to the French nuclear project and received from Joliot's team a report in February 1940 explaining the difference between convergent chain reactions, where the reaction fizzles out, and divergent ones where the whole mass is converted, i.e. a bomb, and the conditions for each to occur. The report also drew Dautry's attention to the very great importance of heavy water in atomic research. Joliot then went to Brussels to meet representatives of the Union Minière du Haut-Katanga to negotiate the purchase of eight tons of yellowcake. In April, the French government was almost persuaded to conduct a secret experiment in the Sahara Desert to see if it was possible to create a bomb but further calculations seemed to indicated that as a result of a sudden rise in temperature the whole nuclear reactor vessel would explode prematurely before a chain reaction could set off a nuclear explosion. The resulting blast would be significant but much less violent and not really worth the cost and effort of such special attention.

In Germany also, researchers had responded to Hahn's paper. On 9 June 1939 a scientist at the KWIC Siegfried Flügge, wrote an article entitled '*Kann der Energieinhalt der Atomkerne technisch nutzbar gemacht werden?*' ('Can the energy content of the atomic nucleus be made technically usable?') after which he was seconded to the Army Ordnance Office. Flügge postulated two possible explosive scenarios. The first would take place when the whole of the fissile material is converted within a thousandth of a second to produce 'an extraordinarily violent explosion' and the second where uranium oxide is mixed with a moderator and the slower conversion takes one-tenth of a second resulting in a less powerful explosion. According to Frisch and Bohr, however, even the former would be far too slow to prevent the container in a reactor bomb from exploding due to the heat buildup before a significant chain reaction could begin.

In Germany the whole nuclear research programme, which at this stage was really very well advanced compared to developments in Britain or the USA, suffered from a lack of enthusiasm shown by the Nazi leaders for science in general. They had an essentially anti-intellectual world view and were suspicious of anything that did not fit into their *Weltanschauung*. It was therefore no surprise when in April 1939 the Ministry of Culture and Education (*Reichserziehungsministerium*, REM) showed little interest in the Perrin and Joliot research papers on fission shown them by the Göttingen physicist Georg Joos. The German army, which might have been more inclined to take an interest when updated by their chemical explosives expert, Paul Harteck of Hamburg University and later by the industrial physicist Nikolaus Riehl, head of a scientific research department in the Auer Company, had no time for nuclear science either. It did not help that none of the nuclear scientists was connected with the National Socialist movement in any way and, as Harteck later explained 'In those days in Germany we got no support for pure science'. Hitler was known to disparage nuclear physics as 'Jew science' which carried with it its own condemnation in the eyes of the administration.[27] As well as the depletion of its scientific ranks by the exodus of – mostly Jewish – nuclear scientists before 1939, the lack of any sort of centralised control of all research, particularly nuclear research, proved to be a serious impediment to the German effort to build an atomic bomb. It was to act as a counterweight to the many advantages that Germany had at the start of the Second World War in terms of its world-class institutions and great depth of scientific expertise. It is noticeable that research and development in other military fields such as rocketry did not suffer in quite the same way perhaps because it had been under development since 1934, was developed at a faster pace and was led by the charismatic Nazi Wernher von Braun.

Of the scientists who stayed and worked for the Nazis, Werner Heisenberg was the most important. He was a scientist of the highest rank as well as being an accomplished classical pianist and a keen student of Greek philosophy who was happy to view science through a philosophical lens. Likening modern physics to the works of Plato he said, 'Modern physics has definitely decided in favour of Plato. In fact the smallest units of matter are not physical objects in the ordinary sense; they are forms, ideas which can be expressed unambiguously only in mathematical language.'[28] He was awarded the Nobel Prize in Physics in 1932 for his work on quantum mechanics which questioned whether sub-atomic particles actually exist objectively and postulated that it was possible to study only our relationship with them rather than the particles themselves.

On 1 April 1935 Arnold Sommerfeld had been asked to choose his successor as the director of the Theoretical Physics Institute at the University of Munich. The process proved to be protracted and contentious because of political differences between the university selection board, the REM and the supporters of the nationalist movement Deutsche Physik. The three candidates were Werner Heisenberg, Peter Debye and Richard Becker, all former students of Sommerfeld. The university favoured Heisenberg as their first choice, but he came under vicious attack by the supporters of Deutsche Physik and by the newspaper of the SS, that called him a 'White Jew' who should be made to 'disappear'. These verbal attacks were taken seriously, as Jews were already being subjected to murder, physical violence and brutal incarceration at the time. For this 'crime', Heinrich Himmler had placed Heisenberg under close surveillance and ordered his henchman, SS-Gruppenführer Reinhard Heydrich, to conduct interrogations in the notorious SS cellars of the Prinz Albert Strasse in Berlin. Heisenberg fought back with an editorial and a personal letter to Himmler in an attempt to get a resolution to this matter and regain his honour. Heisenberg's mother got involved by visiting Himmler's mother, whom she knew as a personal acquaintance, to help resolve the affair. Eventually, Himmler settled things by sending two letters on 21 July 1938. The first was to Heydrich in which Himmler called Heisenberg 'a decent person' whom Germany could not afford to lose or silence. The other letter went to Heisenberg in which Himmler told him that because he had been recommended to him by his family, he would ensure that there would be 'no further attacks on your person' but cautioned Heisenberg against politicising his position as a scientist. This seemed a victory for Heisenberg but that idea was quashed when Wilhelm Müller was given the post of director of the Institute on 1 December 1939. Müller was not a theoretical physicist, being appointed purely for his political affiliations and Nazi credentials.

The politicisation of the education system in Germany had essentially replaced academic tradition and excellence with ideological adherence and trappings, such as membership in National Socialist organisations, like the NSDAP (National Socialist German Workers Party), the NSDDB (National Socialist German University Lecturers League) and the NSDStB (National Socialist German Student League). In 1933 Max Planck, as president of the KWI, had met Adolf Hitler and warned him that forcing Jewish scientists to emigrate would be to the detriment of Germany and benefits of the work of foreign powers but Hitler had responded with a characteristic, spittle-punctuated rant against all Jews in general and Planck was forced to retire without further protest.

The decision by many Jews to leave Germany was the result of the passing of the racial laws in April 1933 which excluded Jews and other non-Aryans from various professions and organisations, including having roles in universities. New laws also cut down the number of Jewish students and those who could practise medicine or law. Prestigious British universities like Oxford and Cambridge quickly hired the most productive German émigré scientists and only the younger, less prominent ones took advantage of a waiver of the US immigration laws to go there. Numerically, it has been estimated that a total of 1,145 university teachers, in all fields, were driven from their posts, which represented about 14 per cent of the higher learning institutional staff members in 1932–3. Out of twenty-six German nuclear physicists cited in the literature before 1933, 50 per cent emigrated.

Paul Harteck was director of the physical chemistry department at the University of Hamburg and an advisor to the Army Ordnance Office (*Heereswaffenamt*, HWA). On 24 April 1939, he and Wilhelm Groth made contact with the Reich Ministry of War (*Reichskriegsministerium*, RKM) to alert them to the potential of military applications of nuclear chain reactions and to make them aware that to make military use of the new science would give Germany 'an unsurpassable advantage over the others'.[29] At the same time, Georg Joos and Wilhelm Hanle notified the REM of the potential military applications of nuclear energy after a presentation by Wilhelm Dames on the use of uranium fission in a *Uranmaschine*, or nuclear reactor, in which the number of fissionable nuclei available in each generation of the chain reaction is carefully controlled to prevent a runaway chain reaction. The goal in a reactor is to get stability by absorbing the exact number of neutrons created in the next generation as the previous generation so that there are same number of fissions going on at all times in a controlled chain reaction as opposed to an uncontrolled chain reaction which would lead to either a reactor explosion or an atomic bomb.

This work came to the attention of Abraham Esau, head of the physics section of the Reich Research Council (*Reichsforschungsrat*, RFR) at the REM, and on 29 April a group organised by Esau, met at the REM to discuss the implications and applications of a sustained nuclear chain reaction. When the industrial firm Auergesellschaft became aware of the interest in uranium, they saw an opportunity to make some money out of a stockpile of uranium they held which was waste material from which they had extracted radium. Its head, Nikolaus Riehl, went to the HWA in July and came away with a contract to supply uranium oxide to the Auergesellschaft plant in Oranienburg, north of Berlin. This led to informal work being carried out at the Georg-August University of Göttingen by a group of physicists and became known informally as the first Uranverein (Uranium Club) and formally as Arbeitsgemeinschaft für Kernphysik. The group was short-lived and was forced to split up in August 1939 when some of its members were called up for military service but it would be a precursor to a second and much more important Uranverein a few months later.

In early 1939, Edwin McMillan ran an experiment at the University of California, Berkeley, bombarding uranium with neutrons using the recently-installed powerful 60in cyclotron. He discovered two new particles as a result of the bombardment: one had the properties of the uranium isotope U-239 and the other he was unable to even guess at. He wrote a paper and called it 'An Unsuccessful Search for Transuranium Elements'. McMillan then collaborated with Philip H. Abelson to try and identify the new particle. They eventually proved beyond all doubt that a new element had been discovered. They published their results in a paper entitled 'Radioactive Element 93' in *Physical Review* on 27 May 1940 and called their new element neptunium. This result quickly spread across the world and would quickly impact on the research being done in Germany by Carl-Friedrich von Weizsäcker. It was to be one of the last US research papers to be released for peer review before a security clampdown on publication of research findings.

At Princeton, Bohr and John Wheeler had provided an explanation of the fission process in their paper 'The Mechanism of Nuclear Fission' published in *Physical Review* of 28 June 1939. Their calculations showed that the heavier an element, i.e. the greater the number of protons in its nucleus, the more likely it is that the electrical charges that cause protons to repel each other will overcome the short-range powerful nuclear force holding them together and this render them unstable. Whereas the short-range force tends to hold the nucleus in a more or less stable spherical shape, the electrical forces are constantly trying to blow the nucleus apart and tend to distort it into an elongated dumbbell shape. If the distortion becomes sufficiently exaggerated the nucleus will undergo fission, i.e. split apart and reform as two lighter elements. Building on the Hahn-Strassmann-Meitner findings, they calculated the magnitude of the energy barrier to be overcome which was related to the relative magnitude of two competing forces and so were able to come up with a single measurement to show which elements were most likely to undergo fission.

Nuclear research was ongoing elsewhere in the US but in an almost leisurely fashion and concentrated more on pure research rather than its potential application. In July 1939, the annual Summer Symposia in Theoretical Physics was held at the University of Michigan. Physicists such as Enrico Fermi and Werner Heisenberg, who attended the symposia, had over the years developed critically important friendships and collaborations through the exchanges of personal letters and manuscripts. This year it was run by Samuel Abraham Goudsmit who had done groundbreaking research on the properties of sub-atomic particles. He was a personal friend of Heisenberg who had been invited to lecture there on topics including uranium fission.

During his stay at Michigan, Heisenberg was seen to have a number of heated conversations with Fermi, with whom he had been close friends for more than ten years since first meeting in Göttingen. Fermi challenged him to explain how 'a decent, honest scientist' could maintain his integrity and self-respect in a country 'where all standards of decency and humanity had been suspended'.[30] Goudsmit joined in and encouraged Heisenberg to stay in the USA. It was suggested that prestigious institutions such as Columbia and Princeton would be only too willing to offer him a job but he was adamant that he could remain aloof from politics and be a force for good back in Germany and anyway it was his firm conviction that 'people must learn to prevent catastrophes [and] not run away from them'. The question of weaponising the uranium research, he said, was moot since 'the war [would be] over long before the first atom bomb [could be] built'.[31] He returned home as one of the very few passengers on the SS *Europa*. Clearly not many people were choosing to cross the Atlantic from west to east at that time.

Back in 1939, working in France, F. Adler and von Halban, in line with Flügge's findings, cautioned that any system containing a critical mass of concentrated U-235 would initiate a chain reaction that would almost instantaneously surge to a staggering 100,000^0 C unless controls, such as cadmium rods, were introduced to absorb many of the free neutrons but Heisenberg had argued that the system could be moderated by ensuring that there was sufficient U-238 in the reactor to stabilise the temperature and result in a 'reactor bomb' that would explode long before all but a small amount of nuclear fuel had the chance to fission but with much less than the power of a true atomic bomb. However, the reactor bomb which became the focus of Heisenberg's research would still require enormous amounts of enriched uranium ore, the production of which was still 'beyond the bounds of possibility [to] affect the course of the war'.[32]

It is important, especially when considering the German research programme, to understand how uranium responds to neutrons of different speeds. It was known that a U-235 nucleus could be split by a neutron collision and that the result would be the emission of two or three extra neutrons as well as enormous amounts of kinetic energy, and calculations suggested that if the uranium was present in sufficient quantity the result would be a chain reaction in which ever more nuclei were sundered and ever more energy was released. We have seen that in a naturally-occurring mixture of U-235 and U-238, fast neutrons will be preferentially absorbed by the U-238 since there is very much more of it. If the neutrons are slowed down then they will be preferentially absorbed by the U-235 but many of them would escape or be absorbed by impurities. This would result in a chain reaction that might be controlled by a moderator to provide power for propulsion for instance, but only if the U-235 is more concentrated, up to maybe 10 or 20 per cent. If such a reactor was set up without a moderator, i.e. with fast neutrons, it might be possible to set up a chain reaction but it would progress only slowly (in scientific terms), although fast enough to raise the temperature of the reaction chamber and cause it to explode. This would be Heisenberg's reactor bomb. Whilst much more powerful than any known explosive, this would not have anywhere like the power of an atomic explosion and it was not yet known what critical mass of uranium would be required.

To set off an atomic explosion would require concentrated U-235 to be bombarded with fast neutrons and this would then allow the chain reaction to progress at a much higher rate, in the region of one millionth of a second, and in this case the chain reaction would happen before the temperature was able to blow the chamber apart and the release of kinetic energy would be almost instantaneous, causing an explosion of incredible power. Making a bomb would mean finding a way of bringing a fissile material up to a critical mass instantaneously and allowing the fast neutrons to do their work.

Having uncovered the mystery of fission, there were a number of theories about how it might be exploited both militarily and in a peaceful context. The difficulty from a practical perspective was that it was inordinately difficult to separate out pure U-235 from uranium oxide that was overwhelmingly made up of U-238. Neutrons fired at a naturally-occurring mixture of uranium oxide would be preferentially absorbed by the U-238 and there would be few available for the U-235 and that would mean no chain reaction and no explosion. The question then became if pure U-235 was too difficult to produce and assuming the U-235 content could be made more concentrated, what percentage of U-235 in a mixture would suffice?

In theory, since U-235 was naturally radioactive, i.e. continuously emitting fast neutrons, a sufficiently large critical mass of pure U-235 might set off a chain reaction of its own volition but that was soon shown not to be the case. It was calculated, however, that an atomic explosion would be likely if two sub-critical masses were brought together to form a super-critical mass in an instant of time such as when they were fired towards each other with high velocity and herein lay the secret of the uranium bomb.

Both of these ideas of the quantity of fissile material required to initiate a chain reaction and the idea that any chain reaction leading to a nuclear explosion would be truncated by a less violent explosion that would occur because of rising temperatures, i.e. a reactor bomb, would continue to point German research down the wrong pathway for the duration of the war. Heisenberg appears to have accepted that a reactor bomb would prevent the chain reaction progressing to the point of a nuclear explosion but that was only the case if slow neutrons were introduced. If fast neutrons were used the chain reaction would be almost instantaneous, resulting in an atomic bomb. While the German approach led them away from making the crucial breakthroughs that would have opened up the way to an atomic bomb, their attitude to the non-military application was clouded by the fear that any chain reaction initiated for whatever purpose could lead to an explosion or a meltdown of the reactor vessel. Even as late as July 1942, a paper published concerning a proposed large-scale reactor stated that 'one must immediately fear that the whole fissioning of the uranium nuclei in an explosive manner in a fraction of a second will occur'.[33]

The first meeting of the second Uranverein was organised by Kurt Diebner in Berlin on 16 September 1939. At the same time the KWIP came under HWA control with Diebner as the administrative director, bringing nuclear research under military control. Responsibilities of the leading scientists were as follows,

- Walther Bothe, Director of the Institut für Physik (Institute for Physics) at the KWI mF in Heidelberg had a team of six physicists working on the measurement of nuclear constants.
- Klaus Clusius, Director of the Institute for Physical Chemistry at the Ludwig Maximilian University of Munich had a team of five looking at isotope separation and heavy water production.
- Kurt Diebner, Director of the HWA *Versuchsstelle* (testing station) in Gottow and director of the RFR experimental

station in Stadtilm in Thuringia, working on the measurement of nuclear constants with six assistants.

- Otto Hahn, Director of the KWIC in Berlin-Dahlem looking for transuranic elements, fission products, isotope separation and measurement of nuclear constants, again with a team of six.
- Paul Harteck, Director of the Physical Chemistry Department of the University of Hamburg, working on heavy water production and isotope production with four others.
- Werner Heisenberg, Director of the Department of Theoretical Physics at the University of Leipzig, with a team of seven or eight researching a uranium machine (*Uranmaschine*), isotope separation and measurement of nuclear constants.
- Hans Kopfermann, Director of the Second Experimental Physics Institute at the Georg-August University of Göttingen, working with two others on isotope separation.
- Nikolaus Riehl, Scientific Director of the Auergesellschaft, with a team of three looking at uranium production.
- Georg Stetter, Director of the Second Physics Institute at the University of Vienna, with six others looking for transuranic elements and measurement of nuclear constants.

Reports from the research conducted were published in *Kernphysikalische Forschungsberichte* (*Research Reports in Nuclear Physics*), an internal publication of the Uranverein. The reports were classified Top Secret and they had very limited distribution. Even the authors were not allowed to keep copies.

It is significant that many of the scientists who worked on nuclear research in Britain and the US were European Jews who had fled Europe in the 1930s to avoid persecution. It was apparent to many of them from the earliest years of Nazi rule that they, along with intellectuals in other fields, would not be permitted to live and pursue their careers free of constraint. Leo Szilard was one such and had moved to England in 1933 to work at St. Bartholomew's Hospital. In England he had helped Frederick Lindemann to establish the Academic Assistance Council to relocate refugee scientists and find them work in universities and scientific institutions in Britain. He subsequently emigrated to the United States on 2 January 1938. There he joined the staff of Columbia University and worked with Walter Zinn on uranium fission research. The nature of his work and the potential for it to be used for military purposes very quickly left him with a sense of foreboding and a conviction that 'the world was headed for grief'.[34] He urged his colleagues to refrain from any further publication of their experiments to prevent them being used by Germany, but it was not until 27 June 1940 that all publication of US scientific articles that had potential military relevance was halted. This would of course have consequences because it could alert the Germans to the fact that prominent atomic scientists who had hitherto been prolific contributors to scientific periodicals were now working on top-secret projects. It was even considered at one point to commission articles on non-secret issues purporting to have been written by Bethe, Fermi and others which might give the impression that they were busy in other fields, just one of the many questionable schemes that popped up with amazing regularity during the war.

Szilard and his colleague Eugene Wigner hit on the idea of co-opting Szilard's old friend Albert Einstein to bring awareness of the potential catastrophic military uses of nuclear fission to world leaders. They sought him on Long Island but could not find his house and no adults they asked seemed to know where it was. They stopped and asked a boy of 'about seven' if he knew where Einstein lived. 'Of course I do,' replied the child and showed them where.[35] On hearing about the recent findings Einstein, who was immersed in his work on relativity and had taken little notice of nuclear research, was surprised when told of the path atomic research was taking and said 'I haven't thought of that at all' but readily agreed to write two letters.[36]

First a letter was sent to the Belgian ambassador to the US for the attention of Elizabeth, Queen of the Belgians, with whom Einstein was acquainted, to warn of the dangers if Hitler invaded Belgium and took control of the open-cast uranium mines owned by the Union Minière du Haut Katanga at Shinkolobwe in the Belgian Congo. The mine had been opened in 1915 primarily for the extraction of uranium ores that contained small amounts of radium but when it had stockpiled what it thought would be sufficient to meet world demand for at least 20 years it had been closed down in 1937 and the mines flooded but the ore had suddenly acquired great importance. The value of Shinkolobwe ore was in its exceptional purity, having up to 75 per cent uranium oxide which was far and away the highest grade ore found anywhere in the world at the time. Shinkolobwe, which means 'a kind of boiled apple that would leave a burn if squeezed', was situated in a pleasant savanna of hills and acacia trees in a region called Katanga, where people had been farming for more than 2,000 years but when it was colonised by the Belgians it became little more than a forced-labour camp where the indigenous people were treated in the most brutal way. Many of them were exposed to lethal doses of radiation. After mining, the ore was crushed in a mill, where water was added to produce a slurry which was then leached with sulphuric acid or an alkaline solution to dissolve the uranium, leaving the remaining rock and other minerals undissolved. The uranium was then separated, filtered and dried to produce yellowcake. The whole area around the mine was heavily polluted and posed a grave risk to the health of people living and working in the area because of the levels of radiation being emitted from the stockpile.

With fellow Hungarian émigré Edward Teller, Szilard and Wigner then composed a letter which they sent to President Roosevelt on 2 August 1939 under Einstein's signature saying 'I believe therefore that it is my duty to bring to your attention the following facts and recommendations'. The letter went on to warn of an 'extremely powerful bomb of a new type' and, since Germany has actually stopped the sale of uranium from the Czechoslovakian mines, urged the President to give 'particular attention to the problem of securing a supply of uranium ore for the United States'. Apart from the Czech source, only two other commercial sources were known, the Belgian Congo and the Eldorado mine at Great Bear Lake in northern Canada.[37] Roosevelt asked the army for an opinion. They concluded that 'there was no basis' for believing that atomic theory would have a military application. They accepted that it 'might have' considerable scientific value but they were not enthusiastic about allocating any funding for research from their own budget

It was thought at this stage that a bomb would require fission by fast neutrons because a chain reaction using slow neutrons might not proceed fast enough to prevent a temperature rise that would blow the pile up prematurely. The word 'pile' was used as shorthand for a nuclear reactor. The crucial question was whether U-235 could fission with fast neutrons in low concentrations but pure U-235 seemed beyond reach. Reports showed that there were currently four methods under development to separate out U-235 from yellowcake. The first was centrifuge separation where the uranium gas is spun rapidly to separate a gaseous mixture of the two isotopes and allow the U-235 to be drawn off and this was tested by Jesse W. Beams at the University of Virginia. It showed promise but there would be no quick and easy solution. The second was the very expensive gaseous diffusion across a porous membrane that would allow the smaller U-235 particles to go through, leaving behind the larger U-238 particles but, despite its cost, it was this method that British researchers were exploring and would later be looked at by John R. Dunning and Harold Urey at Columbia

University in late 1940. The third was liquid thermal diffusion, which used convection currents to carry the lighter U-235 particles to the top and which Abelson pursued at the Carnegie Institution. Lyman J. Briggs, head of the Advisory Committee on Uranium, authorised funding for Enrico Fermi and Leo Szilard's work on fission chain reactions at Columbia University but allocated nothing for gaseous diffusion which was on the point of being closed down.[38] The least promising was electromagnetic diffusion in which the uranium oxide was passed through a magnetic field where the heavier isotope would be deflected relative to the lighter. Alfred O. Nier at the University of Minnesota did some initial work on this process, but it proved to be extremely slow and produced only a miniscule amount of refined U-235.

Fermi thought that a mixture of the right moderator and natural uranium oxide could produce a self-sustaining fission chain reaction, so he and Szilard looked at graphite as a suitable moderator to slow down the neutrons coming from the fission reaction and so reduce the likelihood of them being absorbed by the U-238, increasing the probability of their causing additional fissions in sustaining the chain reaction. A pile containing a large amount of natural uranium could then produce enough secondary neutrons to keep a reaction going. While that might produce a great deal of power and might have military applications in propulsion, it would be far too big for practical application as a bomb. However, enriched U-235 might allow a fast-neutron chain reaction and an extremely powerful bomb, far smaller than any explosive uranium-graphite system, could probably be built.

All during this time most of the work done was in university laboratories by academic scientists funded primarily by private foundations and it was the view of Ernest O. Lawrence, director of the Radiation Laboratory at Berkeley (Rad Lab), that the whole programme needed a shake-up. Lawrence was not a patient man and he launched an energetic campaign to speed up research. Warning Vannevar Bush, the leading scientific adviser in the US at the time, that Germany was undoubtedly pursuing nuclear research with enthusiasm and that the US was likely to be drawn into the war sooner or later, he prevailed upon the National Defense Research Committee (NDRC) to show greater urgency.

By 12 October 1939, Roosevelt had mulled over Einstein's letter and appointed the 'inarticulate and unimpressive'[39] Lyman J. Briggs to head up a new Advisory Committee on Uranium to report back to him on the current state of research on uranium and advise him on what steps needed to be taken. The first meeting of the committee was held nine days later. The committee was well balanced between military representatives who brought scepticism to the table, scientists who brought enthusiasm and waverers who sat on the fence. The first call was for $6,000 to purchase four metric tonnes of pure graphite with further funding to buy 50 tons of uranium ore. Discussion followed resulting in a report submitted to Roosevelt on 1 November which cautioned that 'in view of the fundamental importance and potential military value of nuclear research, adequate support or a thorough investigation of the subject should be provided'. By January 1940 the army and the navy, aware that Roosevelt was 'interested' in the project, had been persuaded to stump up $3,000 each for Briggs' committee that went straight to Fermi at Columbia for the graphite he had called for.

The Uranium Committee held its second meeting on 27 April 1940 at a time there had been an ominous turn of events. In Europe, the Germans had successfully invaded Norway in early April and, as a result, secured control of the Norsk Hydro heavy water plant meaning that they had now obtained a ready source of the substance they were suspected of using as a moderator to achieve a slow-neutron chain reaction. Scientists had now proven definitely that fission occurred only in the U-235 isotope and the University of Virginia, had been successful in enriching a gram of uranium to 10 per cent U-235. The committee still did not make any formal recommendations but agreed that nuclear research should be given every financial support and that steps should be taken to halt all further publications concerning it. Funding from private institutions and the military saw substantial increases.

Meanwhile, the Uranverein team in Germany had set about its multifarious tasks. Erich Bagge had been given the job of testing the idea of using heavy water as a moderator in the fission process. Heisenberg had warned that 'the time needed for the technical development of . . . a reactor depends at the present time mostly on the availability of material, in particular on the production of heavy water'.[40] Heisenberg looked at ways of creating a self-sustaining chain reaction and his first important results had come on 6 December 1939 when he published what became the theoretical foundation for the whole of the German wartime uranium project, 'The Possibility of Technical Energy Production from Uranium Fission' (later referred to as document G-39). It contained these significant sentences:

> According to the data available so far, the uranium fission processes discovered by Hahn and Strassmann can also be used to generate energy on a large scale. The safest way to create a machine capable of doing this is to enrich the isotope Uranium-235. The further the enrichment is driven, the smaller the machine can be built. The enrichment of

> Uranium-235 is the only method with which the volume of the machine can be made small around 1 cu.m. It is also the only method to create explosives that exceed the explosive power of the most powerful explosives to date by several powers of ten. Ordinary uranium without enrichment of Uranium-235 can also be used to generate energy if uranium is combined with another substance that slows down uranium neutrons without absorbing them.[41]

The paper, based on a comprehensive reading of both French and American research to date, detailed how a controlled fission reaction might be contained within a reactor as well as an aside that hinted at how the process might also be utilised in a bomb when 'the whole radiation energy of all the available uranium atoms would be liberated at once', although Heisenberg saw this more as the result of an exploding reactor that had catastrophically gone out of control rather than a specific device designed for explosion.[42] The paper shows that he developed a theory only for the development of a nuclear reactor, laying down no theoretical foundation for developing a nuclear bomb. In all of the forty-one pages of the report, Heisenberg mentioned nuclear explosives on only two and then without any mathematical analysis. The implication is that he had lost immediate interest in a bomb. Both he and von Weizsäcker were concerned more with theoretical problems rather than technical or practical matters.[43] While the paper was a significant piece of work, it was fatally flawed for a number of technical reasons which are beyond the scope of this work to describe but it is an indication of just how 'nebulous, primitive and misconceived' Heisenberg's ideas were at the time.[44] When Sir Frederick Charles Frank, a fluent German speaker who had worked at KWI in the 1930s, interviewed the Uranverein scientist Karl Wirtz after the war as part of British Intelligence investigations into the German uranium project, he was told that Heisenberg gave 'back-of-the-envelope calculations' to a meeting of physicists chaired by Diebner, possibly in January 1940, upon which they 'decided what their strategy for the war period was going to be' implying that they hadn't been particularly well motivated to pursue research into a bomb.[45]

Chapter 3

THE FRISCH–PEIERLS MEMORANDUM

'As a weapon, the super-bomb would be practically irresistible.'
Otto Frisch and Rudolf Peierls, March 1940

In Britain, the most distinguished nuclear scientist in the country at the time, James Chadwick, summed up the rather vague notions of the scientific principles around the atomic bomb. In a letter to the government, he stated that in his view a bomb would probably require 'several tons' of pure uranium, the production of which was almost unthinkable, to react in a thermal-neutron fission process but it was not out of the question that a bomb might also be possible using unrefined uranium. He thought that a reactor bomb might result from the fission of a large quantity, possibly more than a ton, of uranium by the release of thermal (free) neutrons and also considered that a fast neutron chain reaction could lead to a nuclear explosion. His rather hazy views brought comfort to the British government who became convinced that if the problems surrounding the production of an atomic bomb were impossible for Chadwick to resolve then that must also put them well beyond scientists in other countries and so eliminate the possibility of an atomic bomb ever becoming a factor in any imminent war.

Despite Chadwick's inadvertent reassurances, the von Halban and Joliot-Curie results had prompted the eminent British physicist George Paget Thompson to ask the British government to furnish him with a quantity of uranium for research at Imperial College, London. He followed the current line of thinking that U-235 was a suitable fissile material but that huge quantities would be required to build a bomb. That did not mean that the idea was off the table, however. There was no knowing what advances might be made over the coming years in countries like Germany which had a formidable establishment of world-class nuclear scientists. There was every expectation that war with Germany would break out at some point in the not-too-distant future, so Thompson attempted a bluff. He wrote a report designed to be surreptitiously leaked to the Germans in the hope that it might deter them from aggression. This report described a 'successful' experiment in which half a ton of uranium had been successfully exploded in a bomb using a hydrogen and cadmium trigger. It was amateurish at best and it is doubtful if the Germans were impressed.

During the summer of 1939, Winston Churchill had become increasingly alarmed over stories in the British press of a 'super bomb' being developed in Germany that would alter the already precarious balance of power in Europe. He consulted his close friend and scientific adviser Frederick Lindemann. Based on what Lindemann told him, on 5 August Churchill wrote to Sir Kingsley Wood, the Minister for Air, telling him that 'there is no danger that this discovery, however great its scientific interest and perhaps ultimately its practical importance, will lead to results capable of being put into operation on a large scale for several years'.[1] The press again raised the stakes by reporting on a speech made by Hitler made in Danzig on 19 September in which he was reported to have claimed that Germany had 'a secret weapon against which there was no defence'. This subsequently proved to be a mistranslation and a better interpretation would have been, 'we have a weapon with which we [ourselves] could not be attacked', referring to the Luftwaffe, which had no rival of equivalent capability in other countries.[2]

Otto Frisch had left Copenhagen by this time and was working at the University of Birmingham where he was exploring methods of thermal diffusion to separate out U-235 from uranium oxide. He had planned only a short visit, but the German invasion of Denmark had closed the door to his return. Together with Rudolf Peierls he showed how thermal diffusion could be done but to scale up the process to industrial levels to make sufficient U-235 for a bomb would require a gargantuan effort. Peierls was a German Jew who had left Germany in 1929 before moving to Cambridge four years later. He had been a bright child fascinated by science from a very early age. His childhood friends recall him just wandering off in the middle of their play to 'think' quietly and alone about whatever problem was puzzling him at the time. Peierls had actually wanted to be a hands-on engineer but ended up studying physics. His technical aptitude, however, would stand him in good stead later. During his early years he had studied under some of the greatest scientific minds of the age. At the Theoretical Physics Department of Leipzig University, he studied under Heisenberg and then when Heisenberg went off on a world tour he studied under Wolfgang Pauli at the Eidgenössische Technische Hochschule in Zürich. He was awarded a Rockefeller Fellowship and he spent the first half of the year 1932–3 in Rome with Enrico Fermi and the second half in England where he worked at Cambridge with Paul Dirac. Choosing not to return to Germany with its rising tide of Nazism, he went to Manchester to do research with Hans Bethe. When the Second World War broke out, Peierls and his family were classified as enemy aliens but restrictions on them were lifted in early 1940 and they were granted British citizenship but Peierls was still treated with some suspicion and never fully accepted by the British establishment.

In March 1940 Frisch and Peierls would publish their groundbreaking two-part 'Memorandum on the Properties of a Radioactive Super-Bomb' which was the first technical description of how an atomic bomb might actually be built. The memorandum contained the first realistic calculations about the size of the critical mass of fissile material needed for an atomic bomb and crucially revealed that the amount required might be small enough to incorporate into a bomb that could be delivered by an aircraft. Earlier calculations had put the critical mass of uranium oxide to be about 40 tons which could be reduced to 12 tons if materials such as iron or lead were placed around it. Peierls asked what would happen if fast neutrons were fired at a sphere of pure U-235 and his calculations would be literally earth-shattering. This conclusion was the first indication that an atomic bomb might be a practical weapon after all. Such a 'super bomb', they said, would have the explosive power equivalent to 1,000 tons of dynamite and would produce temperatures comparable to those at the centre of the sun. In addition, the explosion would release 'very powerful and dangerous radiations' which would persist for several days. Some of this radioactivity would be carried along with the wind and would spread the contamination; several miles downwind this might kill people.

It was probably rather optimistic of them to say that

> Methods for this separation of isotopes have recently been developed. They are slow and they have not until now been applied to uranium, whose chemical properties give rise to technical difficulties. But these difficulties are by no means insuperable. We have not sufficient experience with large-scale chemical plant to give a reliable estimate of the cost, but it is certainly not prohibitive.

They were on firmer ground when they concluded that:

> It is a property of these super-bombs that there exists a 'critical size' of about one pound. A quantity of separated uranium isotope that exceeds the critical amount is explosive; yet a quantity less than the critical amount is absolutely safe. The bomb would therefore be manufactured in two (or more) parts, each being less than the critical size and in transport all danger of a premature explosion would be avoided if these parts were kept at a distance of a few inches from each other.

Except that they were way out on estimates of the timing which would be shown to be significantly shorter.

> The bomb would be provided with a mechanism that brings the two parts together when the bomb is intended to go off. Once the parts are joined to form a block which exceeds the critical amount, the effect of the penetrating radiation always present in the atmosphere will initiate the explosion within a second or so.
>
> The mechanism which brings the parts of the bomb together must be arranged to work fairly rapidly because of the possibility of the bomb exploding when the critical conditions have only just been reached.

Whilst saying that they did not feel 'competent to discuss the strategic value' of such a bomb, the authors of the memorandum asserted that 'as a weapon, the [bomb] would be practically irresistible' but the violence and extent of the devastation caused by it would probably kill 'large numbers of civilians' which had moral implications. There was no doubt in their minds that Germany would be working on the development of a nuclear bomb and so in order to keep abreast of developments there it was essential, in their view, to monitor both the rate of extraction of uranium from mines in Czechoslovakia and also keep an eye on the activities of one Dr Klaus Clusius whose expertise would be vital to the German programme of U-235 purification. Since there would be no effective way of protecting civilians from the effects of such a bomb, the authors suggested that the best way of countering a German weapon would be for the Western Allies to develop their own as a counter-threat and hopefully maintain a balance of atomic power and deter its use by either side.[3]

In Germany, Heisenberg had identified both heavy water and graphite as suitable moderators and asked Walther Bothe to run tests on graphite given the rarity of heavy water. Bothe's experiments showed that his graphite samples were absorbing so many neutrons that few were left for absorption by the uranium. He attributed the result to impurities in the graphite but when he used what he believed to be a pure sample, the results were even less encouraging. Heisenberg had no reason to doubt Bothe's results, given that he was the recognised authority in the field, and so discarded graphite as a candidate which left only heavy water. In the US Fermi and Szilard had conducted similar experiments and got the same results, but Szilard recalled from earlier work that graphite was commonly impregnated with boron which was introduced during the purification process and boron was vastly more efficient at absorbing neutrons than the graphite. When boron-free graphite was tested its absorption properties were much more in line with requirements and it proved to be perfectly suitable as a moderator. Had Bothe not made his mistake and had Heisenberg not accepted it without question the Germans may well have chosen to use graphite as a cheap and effective moderator in place of heavy water which would have made their work so much easier. When new research in Göttingen conducted by Wilhelm Hanle exposed Bothe's error, Heisenberg realised that graphite would work after all, but by then the whole German stock of pure graphite had been allocated exclusively to their rocket programme.

Chapter 4

HEAVY WATER

Initially, the German programme progressed slowly since only small amounts of uranium oxide and a mere eight litres of heavy water were available to them by the end of 1940. Germany had uranium mines in the Erzgebirge and after the occupation of Czechoslovakia in St. Joachimsthal also but the process for producing pure reactor-uranium, being developed by Riehl at the Auer company, was in its very early stages. Heavy water was the moderator of choice and this had come from the Norsk Hydro (Norsk Hydro-Elektrisk Kvaelstofaktieselskab, Norwegian Nitrogen and Hydroelectric Power Company) ammonia fertiliser plant high above the remote fjords at Vemork near the town of Rjukan in Norway. It had been producing small quantities of heavy water as a by-product since Harold Urey first discovered it in 1934. The fertiliser process used at Hydro involved the hydrolysis of water to produce hydrogen gas which required large amounts of cheap electricity for it to be commercially viable. Hydro had started up a small commercial production plant for heavy water in the cellar of their Vemork factory but it had become useful for scientific research purposes, especially in the field of biochemistry. The plant, set in a deep ravine in the Norwegian mountains, was one of a very few places able to produce heavy water economically simply because of its access to cheap hydroelectric power not only due to the free-flowing streams but also because of the numbers of lakes and water reservoirs which could be regulated to guarantee a steady water supply.[1]

Karl Wirtz, who was one of the scientists working with the Uranverein, was familiar with the Norsk Hydro setup and arranged for representatives of the German chemical giant I.G. Farben, which owned 25 per cent of the Norwegian company, to visit in January 1940. They offered to buy all existing stocks of heavy water and asked if it was possible that production be stepped up, promising that they would purchase the whole output. The management of Norsk Hydro would not agree to the proposal unless they were told what the heavy water was going to be used for. The Norwegians, not unaware of the importance of heavy water in current nuclear research, were uncomfortable with the idea of sending large quantities to Germany and demurred but proved to be more amenable to an approach by Jacques Allier who held a joint position both as a representative of the Banque de Paris et de Pays Bas, which had a 60 per cent financial interest in the Norsk company, and as a lieutenant in the French military intelligence agency, the Deuxième Bureau.

Allier had been briefed by Joliot at the end of 1939 about the crucial importance of heavy water. It was vital, he was told, to find out exactly how much the Norwegian company had in storage and make arrangements to acquire it before the Germans did. On no account, however, were the Vemork management to know about Allier's connection with nuclear research. Allier's mission order read 'Monsieur Jacques Allier is empowered by the Prime Minister to negotiate with the party holding the material that is the object of his mission to assure for France the availability of the largest quantities thereof'.[2]

Travelling under the cover name Freiss, Allier set off on 28 February by way of Amsterdam, Malmö and Stockholm to cover his tracks, accompanied by Captain Muller, Lieutenant Mossé and a civilian, Professor Knall-Demars. In British Cabinet Minutes CAB 126/171 Allier was later quoted as saying that he set off with 'all the paraphernalia of the hero of a spy story with the exception of a false beard'. As a personal friend of its managing director, Axel Aubert, he went to Norsk Hydro in March 1940 to negotiate the purchase of all heavy water from the plant on behalf of the French government. The extreme sensitivity of the mission was clear when von Halban and Kowarski, because of their Austrian and Russian origins (both were by now French nationals), were transferred to the islands of Porquerolles in the Mediterranean and Belle-He off the coast of Brittany until Allier had returned to ensure that there could be no accusations against them if there was a breach of security. Joliot made it clear to Allier that if there was a chance that the heavy water would fall into German hands he was to contaminate it beyond use. Despite Allier being authorised to offer more than ten times its market value, Aubert pledged the whole of its heavy water output to the French government free of charge but left open the option of billing them at some point in the future.

Allier and his team were given 130 litres, the whole of the current stock, which they loaded into twenty-six lightweight aluminium containers, but because they knew that their visit was being monitored by German agents and that they were likely to be intercepted as they tried to get the heavy water out of the country, they planned a subterfuge. It was taken by Aubert himself in his car on a five-hour night journey over icy roads to Oslo and then stored in a house belonging to the French Embassy. Allier now explained to Aubert the vital importance of heavy water, hoping that it might dissuade him from making any future contract with the Germans.

Allier's team booked airline tickets for three consecutive days for two separate flights from Fornebu, one to Amsterdam and the other to Perth, Scotland, both flights due to depart at about the same time each day. On the morning of 12 March, both aircraft were on the tarmac side by side and ready to board. Only Allier and Mossé had come to the airport with large heavy personal luggage and, knowing that their movements were being tracked by Gestapo agents, made a big show of checking in for the Amsterdam flight. They then went out onto the runway towards the aircraft as if to board. Just at that moment a car, which had forced its way through the airfield gate with the help of a British Intelligence agent, Frank Foley, raced onto the tarmac and, hidden from the airport viewing areas, parked between the aircraft both of which by now had their propellers spinning. Ten heavy water canisters packed into suitcases were quickly loaded from the car onto the Perth flight and Allier and Mossé quickly jumped in after them. The aircraft took off immediately. The Oslo-Amsterdam flight they had originally booked on was intercepted by Luftwaffe fighters and diverted to Hamburg where it was searched only to find the heavy suitcases Allier had checked in to be full of gravel.

Meanwhile Muller and Knall-Demars had stayed behind with the remaining canisters and took them on the Edinburgh flight on the following day. As on the previous day, British Customs turned a blind eye when the suitcases were unloaded. All the men met up in Edinburgh and went by train to London then on to Paris where the heavy water was placed in the vaults of the Collège de France. A second mission on 6 April led by a member of Allier's original team, went again to Vemork with a request for more heavy water but were caught in the German invasion of 9 April. The mission was aborted and what samples they had collected were destroyed.[3]

The 'tall, balding, bespectacled'[4] Allier attended a meeting at the Royal Society in London as a representative of the French Ministry of Armaments to discuss the Frisch–Peierls Memorandum on 10 April where he brought the British scientific community up to date on Joliot's work and informed them about the significance of the precious cargo that had been liberated from Vemork and was now stored at the

Collège de France.[5] By this time the whole European landscape had changed with the German invasion and occupation of Denmark and Norway on the previous day. The Norsk plant would soon be taken over and ordered to supply at least one and a half tons of heavy water per year to Germany.[6]

Constant rumours of an imminent German attack against France and the Low Countries posed the question of when and not if it would happen which prompted Sir William Stephenson, head of British Security Coordination, to consider the evacuation of French nuclear scientists and some of their more advanced equipment. With that in mind he sent a team to Paris to identify people and items of value to the British war effort that would need to be brought out if France looked like being occupied. Charles Henry George 'Mad Jack' Howard (aka the 20th Earl of Suffolk, 13th Earl of Berkshire), had run away to sea at age 17, sailing around the world and getting tattoos on both arms, quite an adornment for a member of the peerage at that time. He returned to join the Scots Guards but was then thrown out for his 'wild ways'. Having got a degree in chemistry and pharmacology from Edinburgh University, he used his contacts to get appointed as Liaison Officer for the British Department of Scientific and Industrial Research and since he spoke French fluently and had experience in engineering, it was decided that he should became the British government's contact with the French armaments ministry.

Howard went to France with his secretary and personal assistant Eileen Beryl Morden along with Major Vautier Ardale Golding and his secretary Miss Marguerite Nicolle. Howard was a larger-than-life personality who put many derring-do comic book spies in the shade. Reputed to carry, at all times, two loaded. 45 automatic pistols which he called Oscar and Genevieve, he was accompanied everywhere he went in Paris by a stocky ex-sailor bodyguard with battle scars and a fearsome reputation. In his black wide-brimmed fedora he would be seen driving around the streets of Paris at breakneck speed in a large open-topped car which was not exactly in the best traditions of clandestine operations. One German broadcast was later to refer to Howard as 'The Earl of Suffolk who liked to hide his true activities behind pistol acrobatics in Paris cafes'.[7]

After the German occupation of Denmark and Norway in April, with supplies of heavy water now secure, the Germans further tightened their grip on development of a nuclear bomb by invading and occupying Belgium in May. The Auer company were now ordered to supply German with 60 tons of refined uranium ore a year from their African mines as opposed to the one ton they had provided to date.

A month later France fell to the Germans and by early June their forces were closing in on Joliot and his team of nuclear scientists in Paris.

On 16 May, a few days after the German Panzers broke through the French lines at Sedan, Howard decided that the time had come and requested that Dautry gather together all armament experts and scientists that had been identified as willing to evacuate and at the same time load up as many machine tools as they could get hold of. On the same day, Dautry made arrangements to have all the heavy water along with eight tons of uranium oxide and 1.5 grams of pure radium relocated to Clermont-Ferrand. With German forces on the point of entering Brussels, another batch of eighty-five barrels of uranium oxide was rapidly shipped from Belgium and sent to be stored in a factory near Toulouse in the south of France.

The twenty-six containers of heavy water and documents relating to French nuclear research were placed in the vaults of the Banque de France in Clermont-Ferrand but when German Panzers crossed the Somme and drove south, the bank authorities in Clermont-Ferrand became extremely nervous about having what was clearly sensitive material there and badgered Henri Moreau, who had been given personal responsibility for them by Dautry, to have them transferred to a cell in the Riom prison 10km north of the town. Dautry then contacted Allier and authorised him to take the containers of heavy water to Bordeaux, rendezvous with Howard and get them to England but before that there was what Allier described as a 'regrettable moment of tragi-comedy' when he had to threaten the prison warden with a revolver to force him to have the containers brought up from the prison cells at Riom and loaded into his vehicle.

Howard had arranged through a Mr Irving, the Commercial Attaché at the British Embassy in Paris, to locate the whereabouts of the SS *Broompark*, a relatively new 5,000-ton cargo steamer of the Glasgow-based Denholm Line skippered by Olaf Paulsen, that had been assigned to him for the evacuation of his precious cargo to England. It would be waiting in Bordeaux, he was informed and it was to that port that Howard set out with his lorry-loads of personnel and equipment.

As Paris fell to the Germans, Allier, Joliot, von Halban and Kowarski with their families and the heavy water made their way to Bordeaux along roads choked with refugees and the remnants of the British Expeditionary Force. The British had been reluctant to order a second evacuation of their remaining forces after Dunkirk but when the rapid and total defeat of France was becoming an inevitability, the British War Cabinet first ordered Operation Cycle through Le Havre and then Operation Ariel through Bordeaux to get their men out.

The Royal Navy sent 102 ships and 45 Dutch coasters and the Merchant Navy contributed 129 passenger ships and 141 cargo ships. RAF fighters from five squadrons provided air cover against Luftwaffe attacks as troops and refugees were crammed onto ships and ferried down the long Gironde Estuary out into the Bay of Biscay.

The Germans in Paris had by now found out what had happened to the heavy water containers and had set out to recover them. Travelling by day and night, it was a race against time for the scientists to get to the harbour before German roadblocks were set up. Kowarski and von Halban, both of Jewish origin, were eager to get their families out of France for obvious reasons but Joliot chose to stay to look after Irène who was now in ill health. Allier would stay in France also. Dautry's assistant Paul Bichelonne gave Kowarski a written mission order signed by Paul Reynaud which they were to give to Colonel Mayer at the French Embassy in London authorising them on behalf of the French government to continue research on nuclear energy and make available to the British all the resources Howard was shepherding across the Channel. In this way the continuation of French nuclear research on foreign soil was legitimised. This became the very first example of international nuclear cooperation and also one of the first acts of resistance of the French government. It was later further claimed that the Reynaud letter was 'one of the most important documents in the history of nuclear energy'.[8]

Howard met up with Golding and Nicolle who were waiting at the harbour but, with German troops closing in and German aircraft constantly overhead, there was no sign of the SS *Broompark*. Kowarski and von Halban had eventually arrived, but tensions rose as Howard scoured the harbour for three days trying to get passage on some other ship, then on 13 June, the SS *Broompark* steamed into port. Loading started immediately and afterwards Poulson recorded 101 passengers including 33 French scientists engineers and technicians. There were also a group of Belgian officials, bankers and people connected with the diamond trade including Paul Timbal, managing director of the Antwerp Diamond Bank who had escaped from Belgium entrusted with as much as £3 million worth of diamonds and who had only arrived in Bordeaux on 16 June. Poulson took on 600 tons of cargo including aircraft engines from the French Air Ministry which had arrived from the USA too late for use in the defence of the country and had been left abandoned in railway wagons on the quay. There were also ten motor cars belonging to some of the passengers.

As the SS *Broompark* was leaving the harbour on 19 June another ship loaded with refugees exploded and sank nearby. Minutes later, another ship ahead of them struck one of the many mines that had been dropped in the harbour by Luftwaffe Condor aircraft and sank, and another following was torpedoed. This prompted Howard to consider what would happen to the precious 160 litres of heavy water and the £3 million worth of diamonds if a similar fate should befall the SS *Broompark*. He decided to build a small raft and secure the canisters to it. This would remain on deck so that it would float if the ship was sunk and could be salvaged by survivors in lifeboats. The rest of the journey, however, was uneventful and the grubby collier with its undoubtedly blackened passengers docked at Falmouth on 21 June. Howard informed the authorities that his cargo consisted of:

- 600 tons of machine tools and £3,000,000 worth of diamonds.
- Heavy water from the laboratories of Joliot-Curie in France.
- The entire secret archives of the Ministry of National Education, together with the Under Secretary of State from the Ministry and one other official.
- Two pieces of apparatus of considerable scientific importance.
- All the secret documents belonging to himself and Major Golding from Paris.
- Two new secret machine tools for the manufacture of 20mm Hispano-Suiza cannon.
- Anti-aircraft weapons and twenty-four high-ranking French scientists and technicians together with two pharmacologists and a ballistics expert.[9]

The stocks of heavy water were first stored at Wormwood Scrubs Prison then transferred to Windsor Castle. The diamonds were taken under guard and delivered by Paul Timbal to the Diamond Corporation at 8 Charterhouse Street, London.

The eight tons of uranium oxide that had been placed in the care of physicist Serge Gorodetsky had also been taken to Bordeaux with the heavy water with the aim of getting it to Britain also but by the time he got there the SS *Broompark* was loaded to the gunwales and only the heavy water canisters were loaded. He tried and failed to get passage on any other British-bound vessel willing to take his 130 crates of ore but as a last desperate measure he found a ship willing to take him to Casablanca, from where he was able to transfer it to the mine of the Office Chérifien des Phosphates, located at Khouribga in Morocco. It remained hidden there throughout the war. When the Germans investigated the whereabouts of the uranium, they were told that the ship carrying it and the heavy water had been sunk by a mine in Bordeaux Harbour.

Howard never returned to France. Together with Eileen Beryl Morden and Fred Hards, the other two parts of what he called his 'Holy Trinity', he joined a bomb-disposal unit. They were killed with eleven others on 12 May 1941 while attempting to defuse a booby-trapped German 250kg SC thin-walled general-purpose bomb fitted with a Type 17 clockwork time-delay fuse that had been lying in the 'bomb cemetery' of the Belvedere Marshes for about seven months. His strategy was to closely examine each bomb, dictating notes all the while to Morden, with Hards standing by to assist, on the basis that others would learn from any mistake he might make and not repeat it again. It was a Type 50 anti-handling fuse that had been activated as the Type 17 fuse was removed that caught them out. It was his 35th bomb working in bomb disposal.

Allier was summoned to explain himself before the Vichy Cabinet in August 1940 where he told them the same story that had been told to the Germans, that the uranium ore and the heavy water containers were lying at the bottom of Bordeaux Harbour having been loaded onto a ship bound for England. There erupted a fierce, bad-tempered argument between ministers, some of whom applauded Allier's action and others such as Pierre Laval who had hoped to sell the precious materials to the German occupiers and cursed '*ces salauds d'Anglais*'.

In October 1939, Joliot's team had written several papers some of which were sealed in the vaults of the Académie des Sciences and remained unopened until after the war. These described experiments which showed that only a moderate level of enrichment of U-235 would be sufficient to lead to a chain reaction and that heavy water was a reliable moderator. This had led some scientists to suggest that if France had not fallen to the Germans in June 1940 and the French team had been allowed to develop their theories unhindered, 'the world's first self-sustaining chain reaction could well have been achieved in France'.[10] Joliot later asserted that France had all the elements for making a self-sustaining chain reaction pile in May 1940 but he was adamant that he would never go down that path while Paris was under German occupation.

The Uranverein now had control of much of the world's uranium ore supplies and potential access to significant supplies of heavy water but they lacked a cyclotron particle accelerator. This was remedied from their point of view after the German occupation of Paris when Diebner walked into the Joliot-Curie laboratory and commandeered all their equipment. Joliot returned to Paris from Bordeaux to find his laboratory at the Collège de France occupied by the Germans. He was arrested and placed under the authority of the physicist Wolfgang Gentner whose sympathies were entirely with him and who would go on to protect him throughout the war. Professor Erich Schumann, who represented the German military, allowed Joliot to continue doing research on radioactive isotopes in biology which meant he could avoid any work of a military nature. During the war Joliot would go on to serve in the French Resistance under the alias Jean Pierre Bumont and, in 1941, became the first president of Front national de Lutte pour la Libération et L'indépendance de France, one of the main bodies of resistance to the Nazi occupation. Radio receivers and transmitters for use by the French Resistance would be manufactured in his laboratory throughout the war. In 1944, Irène, who had been unwell for some years due to a compromised immune system having worked with radioactive compounds for most of her adult life, took their children to Porrentruy in Switzerland through secret forest trails. Joliot went into hiding after the Allied D-Day landings in 1944 and supervised the making of Molotov cocktails as the fight for the liberation of Pars drew near.

In June 1945, the Soviets invited all the top atomic scientists to attend a special congress of the Soviet Academy of Sciences. Of the non-Soviet scientists invited only the Joliot-Curies accepted. At the height of the Cold War in 1950, Frédéric would be dismissed from his position as High Commissioner of the Commissariat à l'Energie Atomique at a time when the French nuclear focus changed from peaceful exploitation of nuclear power to making a bomb and a few months later Irène also lost her post as commissioner. The French authorities made no bones about the fact that this was because of their left-wing politics. At an international scientific meeting in Stockholm in July 1953, Irène was denied a hotel room by city innkeepers who disapproved of her political views. She suffered throughout her adult life with tuberculosis and leukaemia and died in 1956 of hepatitis after which she was given a state funeral. Jean Frédéric was to have the same accolade when he died in 1958

Chapter 5

TUBE ALLOYS

In Birmingham, Frisch, 'wrapped up in an overcoat in his freezing rooms',[1] would soon revisit his estimates for critical mass and by 'asking the right questions' revised his estimates of the amount of fissile material required for a bomb. He had initially agreed with the French theory that a temperature rise would destroy any container holding a uranium-neutron bombardment experiment long before enough free neutrons could be liberated to start a chain reaction. The general mood in Britain, much like elsewhere, was that an atomic bomb was only a 'theoretical possibility' that could only become a practical proposition 'in the remote future' and only after a great deal of research and experimentation.

Frisch used the results obtained by Perrin and Peierls and, absorbing the implications of Bohr's work too, decided that as little as a couple of pounds might actually be a critical mass if pure U-235 was used with fast neutrons. The answer was in the rate of production of fast neutrons as a consequence of fission. Looking at the problem from a new angle and applying the results of Bohr's theories, his colleague Peierls went on to calculate that fission would now be speeded up to within a hundred-millionth of a second which would occur before the temperature could rise sufficiently to interfere. Peierls and Frisch, for want of any alternative, took their findings which they had described in their March memorandum to the head of their department, the Australian Professor Marcus Oliphant, currently working on development of the cavity magnetron that would revolutionise radar research. He in turn passed them on to Sir Henry Tizard, the head of the British Aeronautics Research Committee. Tizard was not overly impressed, telling the Cabinet Office that the military significance of the work was 'very low' and asked George Thompson at Imperial College for his opinion. Thompson was sceptical at first, saying that any military use of uranium fission was 'too remote to justify further work in wartime',[2] but when he replicated some of the work in his own laboratory he was 'electrified by the possibilities' opened up by the results and Tizard formed a sub-committee under Thompson to keep track of developments.

Lise Meitner had fled back to Sweden after the German occupation of Denmark and sent a telegram to a friend in England to tell them that she, Niels Bohr, his wife and others were safe. The telegram came into the possession of John Cockcroft at the Cavendish Laboratory in Cambridge who was intrigued by the inclusion of the words 'MAUD RAY KENT' which he inexplicably took to be a cryptic message about radium in connection with a German atom bomb. They in fact referred to an acquaintance of Bohr, Maud Ray, who now lived in Kent. For some reason Thompson now somewhat flippantly renamed his committee the Maud Committee which submitted its first report in October.

When they arrived in London, Kowarski and von Halban got a cool reception because of their alien status. Neither did their research, which centred around using unrefined uranium oxide, seem to be relevant to the British programme which was concentrating on the gaseous diffusion process of U-235 enrichment. They were shuffled off to the Cavendish where they wrote a report for the Maud Committee and got some attention when it mentioned the uranium oxide route for developing a nuclear bomb but the concept remained quite vague. However, the report did touch on the use of plutonium as a possible fissile material, which would not occur to Glenn Seaborg and his team in the US until a few months later.

Thompson was not a nuclear scientist of international standing, nor a particularly renowned experimental physicists and definitely not a talented manager of people. This was amply illustrated in the most stubborn and irrationally xenophobic way when he reduced his committee to the level of farce by denying membership to 'enemy aliens' such as Frisch, Peierls, von Halban and Kowarski who he saw as 'the most unmitigated nuisance'.[3] Despite this, he was considered to be 'well respected by his peers, good at persuading them to talk through their disagreements and quick to size up complicated arguments by summarising the salient points crisply and fairly'.[4] The status of the 'aliens' was further battered when, in late July 1940, every alien in Britain, even those who had been British citizens for many years, received a curt letter with an instruction to report to a police station. Frisch, faced with the prospect of internment on the Isle of Man, was interviewed by the police and it was only with some difficulty that he and other 'alien' nuclear scientists were allowed to remain free to continue their research.

A team of scientists at the Clarendon Laboratory in Oxford headed by Franz Simon set about designing a new gaseous diffusion plant for the production of U-235, having followed Harteck by abandoning the Clusius-Dickel method as prohibitively expensive. Their report was not encouraging but it came along with another report by von Halban and Kowarski suggesting that construction of a nuclear reactor was, after all, quite possible. Theory aside, however, there was still the problem of where the raw materials were going to come from. In September the chemical giant Imperial Chemical Industries (ICI) was awarded a contract to prioritise a supply of highly corrosive uranium hexafluoride and the heavy engineering company Metropolitan-Vickers was given a contract to build a pilot gaseous-diffusion plant at the poison gas manufacturing and storage site in Rhydymwyn, North Wales.

Churchill's top scientific adviser Lindemann was now attending almost every meeting of the Maud Committee which was making ever more encouraging noises about the feasibility of an atomic bomb in the full knowledge that the Germans had probably come to the same conclusion. One of the first Maud decisions was to make known to scientists in the US the whole extent of British nuclear research. The prospect of developing an atomic bomb before the end of the war was now becoming a distinct possibility but it was obvious to the committee that could only happen if Britain was willing to cooperate fully and provide the United States with all the resources and findings of its own scientific institutions to combine with the vast scientific, financial and industrial power of the USA.

Meanwhile, all the activity surrounding nuclear research in the Western world had not gone unnoticed in Moscow. The resident NKVD (People's Commissariat for Internal Affairs) agent in London Anatoli V. Gorsky (aka Anatoli Gromov, code name Vadim) was given a copy of the complete minutes of a 20 September meeting of Maud at which the ICI and Metropolitan-Vickers contracts were agreed. He had acquired this document from his source Listz. This was the Scottish translator and civil servant John Cairncross who had been recruited into the NKVD in 1936 by Norman John (James) Klugman (code name Mayor) a member of the Communist Party of Great Britain (CPGB).

Cairncross, also known to the KGB as The Carelian, was one of the five Cambridge spies recruited by the Soviets in the 1930s, the others being Kim Philby, Guy Burgess, Donald Maclean and Anthony Blunt. Prior to this, Cairncross had also studied at the University of Glasgow and the Sorbonne, Paris. It was a period of economic and political turbulence and this, combined with the rise of far-right extremism, had a considerable impact on many privileged students at the University of Cambridge. Cairncross, however, referred to himself 'a reluctant spy' and one who worked solely in the interests of the Grand Alliance. According to his own brother, Sir Alec Cairncross, he was 'a prickly young man, who was difficult to argue with and resented things rather easily'. There was always the sense of an outsider who was never really accepted by the British establishment despite, or maybe because of, his fierce intellect. Coming from a humble, lower middle-class background in the mining town of Lesmahagow in Lanarkshire, Cairncross had found his social class, accent and mannerisms put him at odds with his privileged Cambridge contemporaries. Heavily reliant on coal mining, his hometown had been particularly badly hit by the 1926 General Strike, the poverty of the Great Depression and far-left political radicalism in the Clyde region. When the Second World War started, Cairncross was recruited as a German translator at Bletchley Park where he worked in Hut 3 of the Enigma Project helping to decode German military communications for the army and intelligence services. Despite his gifts, he was notably inept in many ways. He tended towards clumsiness and was, to the dismay of his Soviet handlers, notoriously unreliable as to keeping appointments and, where he remembered, turning up late. Given a Minox camera to copy documents, he found it so difficult to operate that he actually broke it.

Cairncross worked initially in the Foreign Office and then the Treasury before joining the Cabinet Office, where he worked as a private secretary to Lord Hankey. He had extremely high clearance to British Cabinet Office documents, including minutes of the Committee of Imperial Defence and the War Cabinet, and fed many of them to Gorsky who would pass them over to another NKVD agent Vladimir Borisovich Barkovsky (code name Jerry) who also received secret intelligence from the traitor Donald Maclean. The Moscow NKVD chief, Lavrenti Beria, initially discounted the documents as disinformation designed to panic the Soviet regime into diverting investment into their own nuclear research programme which, as far as he could see, would be a pointless waste when there were more pressing demands on scarce Soviet resources. He did, however, get a second opinion from another physicist, ominously warning him with a blood-chilling threat that he would 'get him into the cellars' if he was poorly advised.[5] On 3 October another serious security leak took place when Cairncross furnished Barkovsky with a seventeen-page report containing information about the size of critical mass required for a bomb and the problems the British were having with the gaseous diffusion method of separating U-235.

By the end of 1940 it had still not been possible to separate out even a small amount of pure U-235 for experimentation but bombardment of U-235 and U-238 mixtures with neutrons of different speeds had convinced the scientists that the whole idea of a bomb was feasible. Simon's work on gaseous diffusion was going well, with ICI manufacturing membranes that would go into the Metropolitan-Vickers uranium refinery plant at Widnes in readiness for the constructing a full-scale pilot plant at Rhydymwyn. Under massive police and army escort, the first consignments of chemicals had arrived earlier in the year but it would not be before the end of 1941 that the first of an estimated 1,000 staff were moved in.

It was becoming ever clearer that an atomic bomb would only become a reality after a great deal of work. Kowarski and von Halban were working on the peaceful application of controlled fission using uranium and heavy water and had made an important breakthrough by confirming what they and Joliot had previously been working on, namely that a slow neutron chain reaction was a practical possibility. The Free French administration in London was keen to have Britain acknowledge the contribution of the French scientists and commit to some sort of agreement that would reward the French government after the war, but the British government was not interested in getting into that conversation and when the US took control of the nuclear programme the French were listened to even less. Peierls was pressing on with research into the gaseous diffusion of uranium through gauzes of very fine mesh and had taken on a new assistant, Klaus Fuchs, whom he had come to know from scientific conferences over recent years. Much will be written about Fuchs in this book and it is therefore worth looking in some detail at his background.

Klaus Emil Julius Fuchs had been radicalised during the early 1930s as a student in Kiel in response to the growing tide of Nazism. Facing arrest for his membership of the German Communist Party (KPD), he fled first to Berlin, then to Paris and finally, with the help of the Religious Society of Friends (Quakers), to Britain on 24 September 1933. Having already been awarded degrees in mathematics and science at Leipzig and Kiel, Fuchs was accepted as a graduate student at the University of Bristol where he completed a doctorate in theoretical physics. Already he had come to the attention of MI5, who categorised him as a 'notorious communist' even though there was no evidence that he had been politically active since arriving in Britain.[6]

From Bristol, Fuchs went on to take up a postgraduate position at Edinburgh University where he started work alongside fellow student Walter Kellermann under Professor Max Born. Fuchs was described by Born as 'weak in appearance but with a powerful brain, taciturn, with a veiled expression which disclosed nothing of his thoughts'. Born's daughter Irène saw something a little different saying that he looked as if he needed 'a warm coat and a good meal'.[7] Born advised him to avoid speaking in public so as not to attract attention to his German accent and also to refrain from all political activity while he was in Britain. Fuchs, however, had quietly strengthened ties with German communist friends in London, especially Jürgen Kuczynski who had arrived in London in July 1936 and had quickly become the political organiser of the KPD in Britain. Fuchs had become one of thirty to forty long-time KPD members, part of a tightly-knit group.[8]

In 1938, Fuchs applied for and was granted an unlimited residence permit but was unable to acquire British citizenship before the war broke out. When Chamberlain declared war on Germany on 3 September 1939, Fuchs lamented that British would limit its efforts to fighting only for the survival of the British Empire and not specifically to defend European democracy against the Nazis. He believed that it was Britain's fundamental aim to provoke a war between Germany and the Soviet Union. Fuchs was categorised along with thousands of other German Jews in Britain as an enemy alien and assessed to find out how much of a threat he was to British security.

Both he and Kellermann were initially put in category C, given 'Refugee from Nazi oppression' status and granted exemption from internment with minimal restrictions on travel and movement. Kuczynski, however, was put in category A and categorised as a serious security risk and interned in Seaton Camp in Devon as an enemy alien on 1 January 1940 but he proved to have powerful allies on both sides of the Atlantic, including the Labour leader Clement Attlee and Lilian Bowes-Lyon, cousin to the Queen, who quickly got him released. Fuchs was not to avoid arrest, however, and received a setback on 11 May when he and Kellermann along with all enemy aliens along the east coast, but surprisingly not including Kuczynski, were ordered to be rounded up and interned. The internees were taken to an armed camp in Huyton, Liverpool where they were each given two blankets, a plate, an enamel mug, but no towels, little soap and no toilet paper. Each man was allowed one bath a month, but the shortage of coal meant there was no hot water. There was a total ban on newspapers, magazines and radios, as well as ordinary mail. When German POWs were brought into the same camp, some of the Jewish 'enemy aliens' who were taunted by them became so traumatised that they were taken away to psychiatric hospitals and one or two even committed suicide

On 4 June, Fuchs and Kellermann were marched through the town with others 'like animals in a zoo' and armed soldiers herded them onto a ship there before the crossing over to another internment camp in Douglas on the Isle of Man. As they disembarked there, crowds of onlookers gawped at them and the local newspaper called them 'ragged creatures'. This strengthened Fuchs' view that the government was intent on appeasing the right-wing press and was pursuing a policy of 'shameful compromise with the Nazis'.[9] Two weeks later, the British government finalised plans to ship thousands of internees under the age of 30 to more remote locations such as Canada, Australia and New Zealand. On the day before he was due to go to Canada, crammed with others into a large, dirty warehouse along the dock, Fuchs wrote to Born 'I have little hope to see Edinburgh again'.

Thirteen hundred refugees boarded the SS *Ettrick* where the portholes had been nailed shut, making the smell almost unbearable. There was a single doorway from the holds where they were kept which would have condemned the vast majority to a watery death if the ship was intercepted by a U-boat in the Atlantic. One of the internees later wrote: 'There is no way out of this trap; the barbed-wire door has been closed and a sentinel guards it. We are truly caught. We are sitting, lying, standing shoulder to shoulder, arm to arm, leg to leg.' Separated from them by a barbed-wire fence, completing the horrific scenario, were 800 German POWs who were also being sent to camps in Canada. Another internee wrote that the sea journey 'was the longest ten days of my life'.[10]

In Edinburgh, Max Born did all he could to campaign for Fuchs' release. He urged Albert Einstein and the mathematician John von Neumann to send reading material to Fuchs and Kellermann who eventually after weeks began receiving physics books and journals. Conditions improved after they arrived in Canada and there was even talk of category C internees being released but the improvement was only temporary. Soon Fuchs and another 600 Jews were separated out and sent to another camp called Camp N near Sherbrooke and Newington, small towns south of Quebec. Here they were put in buildings which were part of a railway depot and found 'dust and soot covering six stinky indoor latrines, five cold-water taps and six lights with bare bulbs . . . train tracks criss-cross the stone floor, which is covered with a layer of foul-smelling mud. There are several large pits in the floor, half filled with stagnant water . . . Everywhere puddles of rainwater have collected. It has either blown in through the broken windowpanes or leaked through the roof.'[11]

In November 1940, when the threat of a German invasion of the British Isles had abated, Alexander Paterson, the British Commissioner of Prisons, visited Camp N with a view to allowing some internees to return to Britain. Just before Christmas, Fuchs and Kellermann were part of a group of 156 internees who boarded the SS *Thysville* and in conditions far superior to any they had experienced since the day of their arrest sailed back to Liverpool. Fuchs would later maintain that he 'felt no bitterness at his internment' acknowledging that it had been 'necessary'.[12]

When Peierls invited Fuchs to join his team at work on the atomic research project in May 1941 Bohr tried to persuade Fuchs to decline the invitation saying that success of the whole nuclear project would mean 'a new concentration of power in the hands of a few capitalists and nationalists'.[13] Fuchs, however, took the view that if the Germans were working on a nuclear bomb then he was justified in helping Britain to match it but his conviction was not total. He asked himself 'would the power be used for good or evil? How great is the danger of Hitler's fascism? How great are the dangers of this frightening weapon in a world torn apart?'

Peierls was told by the British security services that his new assistant's access to nuclear research must be restricted to only those aspects that he needed to know about to fulfil his allocated role. Indeed, Peierls and Frisch themselves had been singled out as potential security risks but out of dire necessity they were, albeit reluctantly, seconded to one of the Maud projects. Peierls argued that such a restriction placed upon Fuchs' work was quite impractical and had the condition lifted, giving Fuchs clearance for all Maud findings. The new man gained a reputation for erudition but was seen as reserved to the point of taciturnity.

The whole Western effort to build an atomic bomb turned a corner with two Maud reports of 15 July 1941. One was entitled 'Use of Uranium for a Bomb' and the other 'Use of Uranium as a Source of Power'. The authors were convinced that 'release of atomic energy on a large scale is possible and that conditions can be chosen which would make it a very powerful weapon of war'.[14] It went on to say 'In spite of . . . very large expenditure we consider that the destructive effect, both material and moral, is so great that every effort should be made to produce bombs of this kind' and 'the material for the first bomb could be ready by the end of 1943'.

Attention now turned to devising a mechanism that would translate bomb theory into a practical application. In other words, to design an actual bomb. It was known that below a certain critical mass pure U-235 would not explode so the answer seemed to be to find a way of rapidly bringing together two sub-critical masses to create a critical mass which would then explode in a chain reaction. The Maud report said:

> In order to achieve the greatest efficiency in an explosion of this type, it is necessary to bring the two halves together at high velocity and it is proposed to do this by firing them together with charges of ordinary explosive in a form of a double gun. The weight of this gun will of course greatly exceed the weight of the bomb itself, but should not be more than 1 ton and it would certainly be within the carrying capacity of a modern bomber. It is suggested that the bomb (contained in the gun) should be dropped by parachute and the gun should be fired by means of a percussion device when it hits the ground. The time of drop can be made long enough to allow the aeroplane to escape from the danger zone and as this is very large, great accuracy of aim is not required.
>
> Although the cost per lb. of this explosive is so great it compares very favourably with ordinary explosives when reckoned in terms of energy released and damage done. It is, in fact considerably cheaper, but the points which we regard as of overwhelming importance are the concentrated destruction which it would produce, the large moral effect and the saving in air effort the use of this substance would allow, as compared with bombing with ordinary explosives.

The report noted that Americans seemed to be working on the uranium problem from the aspect of uranium as a source of power rather than the production of a bomb and concluded by saying:

> (i) The committee considers that the scheme for a uranium bomb is practicable and likely to lead to decisive results in the war.
> (ii) It recommends that this work be continued on the highest priority and on the increasing scale necessary to obtain the weapon in the shortest possible time.
> (iii) That the present collaboration with America should be continued and extended especially in the region of experimental work.

Churchill reviewed the report and concurred with Lindemann that work should begin on an isotope-separation plant immediately but in Britain and not the USA. His view was conditioned by a conviction that the country that held the lead in nuclear research would 'be able to dictate terms to the rest of the world' and 'however much [he] may trust [his] neighbour . . . [he] was very much averse to putting [himself] completely at his mercy'.[15] The closest cooperation between Britain and the US, however, was vitally important and great efforts were now made to bring both countries together for further research. Churchill would later say that, in his opinion, it was the British clear analysis, strong theory, practical programme, its sense of urgency and its confidence in the Maud report that changed US opinion and supercharged the project from that moment on. Marcus Oliphant was despatched across the Atlantic in an unheated bomber ostensibly to liaise on radar development but also to galvanise the US scientists into moving their nuclear research up a gear. Oliphant, with his tight curls of greying hair, gold-rimmed circular spectacles and neat three-piece suit, often resembled a lovable schoolmaster but in his younger days had been called 'hot-headed and impulsive'.[16] He lobbied energetically in the United States in what was almost a single-handed effort to shake things up and then did the same at home in Britain, urging officials to act quickly and effectively on the Maud report's conclusions.

Accepting that 'a vigorous British atomic project was now justified', Churchill endorsed the setting up of the Tube Alloys Directorate in November 1941, officially a branch of the Department of Scientific and Industrial Research. Despite the confident tone of the Maud report, a great deal of fundamental research remained to be done and an important early function of Tube Alloys was to co-ordinate and expand this work. The universities of Cambridge, Birmingham, Liverpool, Oxford, King's College, London and Imperial College, London were all involved, with Bristol and Manchester universities soon brought in. Unfortunately, many of the leading British scientists were already employed in other war work but the presence of many well-qualified refugees from Germany and occupied Europe meant that meaningful work could be undertaken. It is worth noting the ironic contribution to Tube Alloys made by émigré scientists who were available to counter the German threat who were in Britain as a result of Germany's own racial policies. Tube Alloys pursued two main avenues of research, isotope separation and reactor development, but it soon became clear that swift progress would depend on the enormous industrial and scientific potential of the United States.

The Maud report had been discussed on 16 September at a meeting of the Defence Services Panel of the Scientific Advisory Committee chaired by Lord Hankey whose private secretary happened to be John Cairncross. This ensured that the details of the meeting quickly came to the attention of the Soviet spymaster in London, Anatoly Gorsky, and were reported to Moscow. The Soviet Union now knew beyond all doubt that Britain was committed to the construction of an atomic bomb, but Moscow was to receive an unexpected bonus when Klaus Fuchs bumped into his old friend Jürgen Kuczynski at the house of the Free German League of Culture on Upper Park Road in Belsize Park, London in April 1941.

Kuczynski was a German-Jewish economist and journalist who had come to England in 1936 after spending some time in Moscow where it is thought that he was recruited by the GRU (Soviet Military Intelligence). At some point before the start of the war it is believed that he was also recruited by one of the US intelligence agencies which might explain why he had been treated so leniently during the roundup of 'aliens'. Another reason was his acceptance by the British establishment, even up as far as Churchill himself, for his undoubted intellect, experience and worldly wisdom. He may also have been an unofficial contact between Western and the Soviet governments because he was allowed total freedom to pursue his communist activities in Britain and was known to be a regular visitor to the Soviet Embassy in Kensington Palace Gardens. His activity as a Soviet-US double agent was probably not known to MI5, however, who continued to see him as a person of interest due to his loud and persistent assertion that 'Europe must be prevented above all from falling under English domination'.[17]

It had been the German invasion of the Soviet Union in June 1941 that convinced Fuchs that the Soviets, now allies of Britain, should be given access to British research on atomic weapons and he unburned himself of his feelings to Kuczynski who arranged, through the Soviet Ambassador Ivan Maisky, for him to meet Simon Davidovich Kremer (code name Alexander) who was well known to MI5 as a Soviet Intelligence agent. The two were introduced and continued to meet regularly but always at a different location well away from the embassy. Fuchs claimed that, at this stage, he 'did not have the slightest doubt that Soviet foreign policy was correct and . . . was convinced that the Western allies were [doing their utmost] to see that the Soviet Union and Germany would completely exhaust themselves in their struggle to the death'.[18] However, as a sop to his conscience, he decided only to give the Soviets information about the work with which he was personally involved. He did not have security clearance to see top secret documents but that was immaterial since much of what was in those papers was the results of his and Peierls' research in the first place. He passed to Kremer a large notebook containing details of the British Tube Alloys project along with copies of papers he had written and advised him that work had started on the building of a bomb in both Britain and the USA. Fuchs became an invaluable source for Soviet Intelligence and was given the code name Rest and later Charles.

Fuchs' handler in Britain, Kremer, was recalled to Moscow and he was left with no contact but Kuczynski soon remedied that by handing him over to his sister Ursula (code name Sonja) whom Fuchs had known already for several months. Born in Berlin in 1908, one of five sisters, she had been radicalised at a young age by the poverty and deprivation she saw. Her globetrotting started early and took her to New York where she worked in a bookshop between December 1928 and August 1929 before returning to German to marry Rudolf Hamburger, like her an active member of the KPD. Hamburger, an architect, went to Shanghai to work in 1930 and took Ursula along with him. There she was recruited to the GRU by the German expatriate journalist and Soviet spy Richard Sorge, who became her lover. She went to the Soviet Union on an espionage training course for seven months even though she had a small child whom she left behind in the care of her parents-in-law who lived in Czechoslovakia. 1934 found her based in Shenyang in Manchuria which was under Japanese occupation, where she excelled as a radio operator. Thereafter she spent three years with Sorge in Poland setting up resistance groups that became active after the German occupation. It was at this time that she was made a colonel in the Red Army and awarded the Order of the Red Banner for her espionage work in China.

Between 1938 and December 1940, Ursula was in Switzerland working with the Lucy spy ring that fed intelligence from inside Germany to the Soviet Union. During this time her first marriage ended and she married Len Beurton, a fellow communist who had fought with the British Battalion of the International Brigade in Spain. This gave her British citizenship, allowing her in February 1941 to move to England with her new husband, settling first in the Oxfordshire village of Glympton and then Kidlington which was close to where her parents lived having emigrated from Germany in 1933. The village just happened to be located close to the UK's Harwell Atomic Research Centre and Blenheim Palace where much of British Intelligence had been relocated at the start of the war. She was interviewed on a number of occasions by MI5 but managed always to greet them at her house disarmingly immersed in domesticity wearing an apron, looking after her children and dusting flour off her hands.

Beurton and Fuchs met twice in September 1941 in Banbury when Fuchs handed over documents and they then walked together around the quiet Oxfordshire countryside for a while, occasionally arm-in-arm to look as inconspicuous as possible. Fuchs would pass across to her secret material which was then enciphered and radioed to Moscow from Beurton's cottage in the heart of England. Ursula, who kept her true identity and address from Fuchs, found him 'calm, thoughtful, tactful and cultured'.[19] After 18 June 1942, when Fuchs signed the Official Secrets Act, he was granted access to all 'matters of relevance' concerning Peierls' team and others in the Tube Alloys programme.[20] At this time Fuchs would have been working on gaseous diffusion and he handed over all he knew on that topic. He also passed on what was the official British understanding that Germany was not making much progress on its own bomb and most unlikely to have one before the end of the war. This allowed the Soviets to hold back on excessive expenditure for their own research although Fuchs did also suggest that work towards a bomb in the USA was progressing apace. After a pause of more than a year, Fuchs and Beurton met a further five times until 22 November 1943 when Fuchs got his US visa having been granted British citizenship on 31 July 1942. His intelligence now was going to the NKVD and not the GRU. The two agencies had struck an uneasy truce where atomic intelligence was concerned. Ursula now briefed Fuchs on how to get in touch with his new contact (code name Raymond) in New York on the first Sunday in February.

Meanwhile, Beurton was running another agent, Melita Norwood, née Sirnis (code name Tina), the daughter of a Baltic émigré. Sirnis had been active as a Soviet agent since 1934 when she was recruited by Andrew Rothstein along with her mother Gertrude and her sister Gerty while living at the notorious Lawn Road Flats where the NKVD recruiter Arnold Deutsch and René Kuczynski, the father of Jürgen and Ursula, also lived. Gertrude trained Moscow recruits in the use of secret transmitters at the family home. Melita became a British citizen by marrying a Russian-Jewish mathematics teacher Hilary Norwood who had changed his name from Nussbaum.

As Britain started to re-arm, the Non-Ferrous Metals Research Association, where Melita Norwood worked, started to play an important role in new weapon design for the British armed forces at the Woolwich Arsenal. Norwood copied confidential documents relating to metallurgy and passed them onto her Soviet contacts. In 1937, MI5 began to investigate Percy Glading, a former Royal Navy gun examiner who worked at Woolwich and was suspected of espionage. Glading and two accomplices were caught in possession of secret documents, tried and imprisoned but Norwood, whose name appeared in Glading's secret notebooks, was not rounded up with them. MI5 had opened a file on her at the time and began surveillance at her home but for reasons unknown no further action was taken.

In an article for *The Observer* on 3 July 2005, David Rose described Norwood's war work. During the war, he said, she worked as secretary to G.L. Bailey, the chairman of the British non-Ferrous Metals Research Association, a company that had become associated with the Manhattan Project. In this role, Norwood gained access to documents that her Moscow handlers described as being 'of great interest and a valuable contribution to the development of the work in [the atomic] field'. Norwood described how easy it had been for her; 'Sometimes, if I was typing something, I typed an extra copy. I typed up the minutes of meetings. I passed on copies by arrangement, left them somewhere pre-arranged or met somebody and handed them over.' She was never challenged when leaving the office with classified documents. 'There might be spare copies on the shelves', she said, 'I just helped myself.'

Norwood was variously described as 'both the most important British female agent in KGB history and the longest-serving of all Soviet spies in Britain' and a 'committed, reliable and disciplined agent, striving to be of the utmost assistance'. British Intelligence files described her as 'a rather dotty old lady who did a spot of amateur spying'. When she was exposed later in 1999, the British government claimed that the information she supplied to the Soviets was 'of little significance' because the Soviets had not used British designs for their reactors. This was quite untrue; the Chelyabinsk reactor was copied from the British / Canadian Chalk River design. The British government went on to announced that, at the age of 87, there was no point in prosecuting Norwood which at least spared them the embarrassment that a prosecution might bring.

Chapter 6

THE US OFFICE OF SCIENTIFIC RESEARCH

'A single lurid fear brought the American decision to undertake the vast effort and expense required to build the atomic bomb – the fear that Hitler's Germany would do it first.'

Thomas Powers, *Heisenberg's War: The Secret History of the German Bomb.*

The first attempt to interest the US military in nuclear research had been on 17 March 1939 at the time of the German annexation of Czechoslovakia. Enrico Fermi managed to get a meeting with Admiral Stanford C. Hooper, Director of the Technical Division Office of the Chief of Naval Operations. Hooper was already aware of the latest scientific developments and was very interested in the potential use of atomic power for propulsion, especially in submarines, but he had little interest in a bomb. When Szilard had a follow-up meeting, even though the Navy was following its own programme of research into uranium isotope separation he was told that it would be 'almost impossible' to reach an agreement that 'would be really helpful'.[1] In Hooper's mind, Fermi's estimates of the costs involved made the whole concept of a bomb no more than 'of academic interest' and the Navy had little enthusiasm for the idea.

It is doubtful if Fermi himself was actually taking the idea of a bomb very seriously. He had, however, been thinking about the use of unrefined ores with a U-235 content of less than 1 per cent but there was a chance that the theory behind a bomb could be taken more seriously if pure U-235 could be manufactured. At the time, the only method of purifying it from natural ores involved the Clusius and Dickel process of thermal diffusion in which the ore was heated to a point where it became a gas. The idea behind this process was to have two tubes, one inside the other, with one heated and one cooled. The slightly lighter U-235 isotope would move towards the heated tube and rise while the heavier U-238 isotope would move towards the cooler tube and descend. The U-235 would then be collected at the top of the apparatus. The process, unfortunately, involved working with a noxious substance, uranium hexafluoride, which 'corrodes just about anything it comes into contact with'.[2] In one important way it was an ideal substance to use because it could be converted into a gas at much lower temperatures than other uranium compounds but few people had any experience of working with it. Even so, Fermi was still not convinced that a bomb was a practical proposition, saying in early 1940, 'Even if reaction chains can occur there will remain too short and too rare for there to be any question of an explosion'.[3]

The US was not yet at war and most scientists were unwilling to invest either time or effort in thinking about weapons of mass destruction. They were certainly aware of the theoretical possibilities of a 'super bomb' but it remained in the realms of hypothesis with no relevance to their everyday work. Many senior scientists and engineers who prepared the reports that served as the basis for government policy decisions did not take the time to learn the essential facts of nuclear power and so failed to grasp its significance.

With war in Europe a reality and Japan making threatening moves in the Far East, in the spring of 1940 Vannevar Bush, a man 'who looked like a beardless Uncle Sam'[4] adjudged that the country's infrastructure of science agencies would not serve the needs of the coming war. For instance, there was no government agency with the authority and means to proceed with the development of new weapons and technology or even assess the technological deficiencies of the military and say what might be required. On 12 June 1940, Bush persuaded Roosevelt to approve the establishment of the National Defence Research Committee (NDRC) with Bush himself at its head to mobilise science for national defence and to serve as a liaison between scientists and the military. This effectively reorganised the Uranium Committee into a scientific body and eliminated all direct military membership. However, despite his position, Bush had not kept up with the latest developments. He was still not terribly *au fait* with the science behind the bomb project and, seeing that the NDRC could no longer rely on the military for funds, he was dubious about allocating large sums of money to research into such an esoteric field. He feared that 'wild notions' about the destructive power of an atomic bomb would spark public unrest and he was reluctant to 'take any bold steps' and, in his own words, he was 'puzzled' over what to do.

Another issue crossing the desk of Briggs at the Advisory Committee on Uranium in April 1940 was that it was clear that the Germans had become interested in the Norwegian heavy water plant at Vemork which was a strong indication that they were likely to be researching atomic power. The other indicator that would support the idea of a viable German atomic programme was any increased interest in uranium ore. At that time it was known that large stocks of very high-grade ore were stockpiled at Shinkolobwe in the Katanga province of the Belgian Congo from where it was being shipped in small quantities to a Union Minière facility in Belgium for use by the Joliot-Curies at the Collège de France. The Belgians had agreed to supply the French with eight tons of high-grade uranium oxide, having been offered a half share in patents the French had drawn up regarding the principle of a nuclear reactor. When Belgium was occupied by German troops in May 1940, however, the prospect of all the ore being utilised for German nuclear power became clear. Despite calls from Brussels for the governor-general of the colony, Pierre Ryckmans, to remain neutral he firmly declared the colony's support for the Allies and ordered the internment of all German nationals of military age. The German occupiers of Belgium, however, set up the Kolonial Politisches Büro to facilitate arrangements for German trade with the Congo, especially for its rubber, diamonds and minerals including uranium.

It was known that Edgar Sengier, the managing director of the Union Minière du Haut Katanga mining operation in the Congo, was in New York and had previously authorised a shipment of 120 grams of radium, worth nearly $2 million to the US but little or none was sent before the German invasion of Belgium. It was really the radium that the Belgians were interested in at Shinkolobwe but this was mined along with the uranium ores and separated out later. Sachs and Urey went to see Sengier in early June 1940 but, while he was willing to give them information about the size of stocks of uranium ore in the Congo, he would not agree to ship it to the USA. This left the USA reliant on Canadian sources but the Eldorado mine there would probably not be able to produce and ship ore for at least another year and, in any case, the ore there was of much lower purity than the Shinkolobwe product. Sengier, meanwhile, concerned that his stocks might fall into German hands and, no doubt realising that his ore would become ever more valuable if the US entered the war at some point, had ordered the shipment to New York of approximately 1,200 tons of high-grade ore from the Shinkolobwe stockpile in the Congo. The shipped ore was then stored in a warehouse in Port Richmond on Staten Island at a subsidiary of Union Minière called the African Metals Corporation (Afrimet) but Sengier, a 'polished man, somewhat stout, with pale skin, white hands, a fringe of white hair and a short silver moustache', was in no hurry to cash in his assets and made no effort to inform the American government officials at the time. Soon afterwards, Tizard had tried to buy the whole Katanga stockpile of uranium ore for Britain but was told that it had already been shipped it to the USA. Despite the lack of cooperation from Sengier, the Americans, quite unaware that Congolese ore was already in the USA, decided to keep a close eye on the Shinkolobwe mine, primarily to monitor any shipments that might be going to Europe.

With increasing enthusiasm in the USA for atomic research and a realisation that the British had already made significant advances, in February 1941 Roosevelt sent James Bryant Conant, the president of Harvard, to London to set up a liaison office between the NDRC and the British government. Conant met Churchill and Lindemann who drew him into a discussion about an atomic bomb which was 'the first [Conant] had heard about even the remote possibility of a bomb'.[5] Kenneth Bainbridge, whom Conant had taken along with him, was allowed to attend a meeting of the Maud Committee and was surprised to find that the British had 'a very good idea' of how a bomb might be assembled but lacked the sort of industrial or financial resources necessary for the large-scale production of either heavy water or graphite, the proposed moderators. Seeing how far ahead in research towards a bomb the British were, Bainbridge had every reason to suspect that the Germans were at least as far advanced, if not more so, and urged that the US send someone on permanent assignment to join the Maud Committee to keep abreast of British developments but Bush was still struggling with his doubts about getting too involved with a bomb that threatened to consume vast resources with no clear prospect of success before the war ended. He needed 'an energetic but dispassionate review' and brought in the Nobel Prize-winning physicist Arthur Holly Compton to lead it.

Compton came up with a report on 17 May 1941 which looked at critical mass issues which would determine the size of reactor required to initiate a chain reaction but ignored any practical steps that might lead to bomb assembly mechanisms. It concentrated primarily on the possibility of nuclear fission being useful as a power source for ships and submarines but peripherally included a discussion of the possible munition aspects of the research. The report concluded, as a general principle, that no atomic bomb could be ready before 1945. Conant reinforced Bush's scepticism by saying that 'only efforts which were likely to yield results within . . . a year or two were worthy of serious consideration', which seemed to rule out any work on an atomic bomb.[6] The whole US programme had arrived at a crucial moment. Compton clashed with Conant and Bush who, he believed, were 'very close to dropping the whole fission project from the war programme [altogether]'.[7]

On 2 July 1941, the NDRC considered a request from Briggs for $140,000 to purchase uranium metal and pure graphite but its members were not very supportive and could not agree that funding for nuclear research should be given priority over other areas. They noted, however, the grave danger that German nuclear research might prove successful and therefore hedged their bets by urging caution which became the order of the day. They agreed to a 'careful, but not elaborate or expensive program'.[8] Bush knew he could not risk overlooking what might prove to be an important development and so decided to take a fresh approach by calling in William David Coolidge of General Electric to replace Compton with whom he had not established any sort of easy working relationship. Representatives of Bell Laboratories and Westinghouse were co-opted also. As a result, nearly $500,000 was allocated for work to be split amongst Columbia, Harvard, Princeton, the University of Minnesota, the Standard Oil Development Company, Iowa State College, Cornell, the University of Chicago, Johns Hopkins, the Carnegie Institution of Washington, the University of California, the University of Virginia, the Bureau of Standards and the Department of Agriculture.

However, the nuclear research that was being undertaken across US campuses ostensibly to look at atomic power as an energy source could not avoid stumbling across lines of inquiry that led to a bomb. The Princeton physicist Louis Turner had contacted Szilard to elicit his opinion in May 1940 on whether or not to publish a paper he had written on uranium fission. He was concerned that his research findings might be of a sensitive nature and publication might not be in the national interest. In his paper he had postulated that even if a mixture of U-235 and U-238 was bombarded by slow neutrons some of the U-238 would still absorb neutrons and become a new material which had no name as yet, but which would probably be very unstable and might decay into another new substance with 94 protons which he called eka-osmium. Either or both, of these new materials, he thought, might be even better candidates for fission than U-235.

When McMillan and Abelson at Berkeley had published their findings on plutonium, von Weizsäcker's mind was already fertile ground for its reception. He persuaded Heisenberg to look at the idea which resulted in a paper in July 1940 entitled 'A possibility of Gaining Energy from U-238'. If the U-238 absorbed a fast neutron then it might produce an unstable isotope U-239 that would quickly decay into an element with atomic weight 93 which McMillan and Abelson had called neptunium. This element, heavier than uranium, that did not occur naturally anywhere on earth, promised in theory to be even more fissile than U-235 but it turned out that it decayed quite quickly and so appeared to be of no practical use as an explosive. When von Weizsäcker read follow-up papers by McMillan and Abelson showing that neptunium decayed into plutonium, atomic weight 94, which was more stable, he realised that this would be produced in pure form and so would be a much better candidate for fission that the elusive U-235. The essential feature of plutonium was that its manufacture did not require any expensive and difficult separating out of the uranium isotopes. The unseparated ore would work just fine in a reactor specifically designed for plutonium production. The mechanism of generation of plutonium involved several stages. First by capturing a neutron, U-238 becomes U-239 (92 protons and 147 neutrons). This then rapidly changes by beta radiation into neptunium, Np-239 (93 protons and 146 neutrons), which decays through beta radiation, after three days on average, into a new substance, plutonium, Pu-239 (94 protons and 145 neutrons). This rekindled von Weizsäcker's enthusiasm for the creation of a bomb, an idea he was quick to bring to the attention of the HWA even though it did not fit in with all other German research that was still concentrating on the heavy water approach using U-235 and slow neutrons.

McMillan and Abelson's paper had made the whole scientific world aware that a uranium reactor could now produce new fissionable material suitable for making an atomic bomb. This was certainly not overlooked by Heisenberg's team in Germany but they remained convinced that the way forward was through using a U-235/U-238 mixture with heavy water as a moderator. In Britain, Turner's theory got little traction either, with scientists there concentrating on the purification of U-235 through gaseous diffusion even though Egon Bretscher and Norman Feather at the Cavendish Laboratory had realised that a slow neutron reactor fuelled with uranium would theoretically produce substantial amounts of plutonium as a by-product. And this could be readily split by both slow and fast neutrons with the added advantage of being chemically different from uranium and consequently easier to separate out. Glen Seaborg built on Turner's theories and had shown how plutonium was 1.7 times more likely than U-235 to undergo fission with slow neutrons and since pure U-235 was so difficult to obtain, developing a fission bomb using plutonium became a promising avenue of investigation but plutonium lay on the other side of uranium fission which itself was still a long way from realisation.

The NDRC was well suited to manage the laboratory research into fission, but it did not have the expertise to oversee the sort of large-scale engineering programme that would be essential to design and build facilities to feed and operate an atomic reactor. On 28 June 1941, Roosevelt issued Executive Order 8807 authorising the setting up of a new body, the Office of Scientific Research and Development (OSRD), under Bush and reporting directly to him. It would be the centre for the mobilisation of the scientific personnel of the nation and coordinate activities of the Departments of War and Navy as well as other government agencies. Bush would advise the President on scientific matters concerning national defence and be a liaison between Allied nations. He reassigned the Uranium Committee to the OSRD and established a planning board to study the engineering of facilities for the production of atomic weapons. The OSRD would go on to play a crucial role in the development of new advanced weaponry, including the atomic bomb.

The NDRC, now under Conant, became an advisory body to the OSRD and the uranium section of OSRD was now referred to simply as S-1 (Section-1) for security reasons. When it reported in July, however, it hardly touched on any advances in plutonium research or its potential application in a bomb. Many members of the Uranium Committee were still under the impression that all nuclear research was geared towards 'making a power source for submarines'.[9] At this time, as a result of the information exchange programme, Bush also received an official copy of the Maud report of July 1941 which clearly showed that Britain did not have anything like the resources necessary to build an atomic bomb on its own. As early as 1940, Britain had made tentative overtures to the Canadian government to allow Kowarski and von Halban to transfer their research work to Canada where resources, especially cheap hydro-electric power, would hugely facilitate development.

These two men had built on the research done with Joliot in Paris despite having been greeted rather coolly in Britain when they had arrived on 21 June with their precious cargo of heavy water. The French approach using natural uranium ore with heavy water as a moderator was considered to be of secondary importance and neither of the two men were even allowed to join the Maud Committee although they were permitted to work at the Cavendish. The British had been much more interested in working with pure U-235 they hoped to produce by gaseous diffusion but by the end of 1941 Kowarski and von Halban's work in Cambridge had clearly demonstrated that a slow neutron chain reaction using uranium oxide and heavy water was eminently possible. Chadwick urged that publication of their research paper 'Technological Aspects of Nuclear Chain Reactions used as a source of Power' be withheld from publication for security reasons.

Oliphant had come to the US in late August 1941 to see why there had been so little response from the US to the Maud report. He was anxious to divert US research into development of a bomb rather than its current preference for energy production. Apparently, the report had remained locked in Briggs' desk and was not widely circulated. Oliphant found Briggs to be 'distressingly dull, bumbling and obsessed with secrecy'[10] and immediately took it upon himself to remedy the situation by persuading Bush to send two members of the Uranium Committee, George B. Pegram and Harold Urey, to Britain in October 1941 so that they could be made aware of exactly where the British were with their research.

Although he had received an unofficial version, it was not until 9 October, three weeks after it had been seen in Moscow, that Bush put the Maud report on Roosevelt's desk. The President now had in front of him specific plans for building a bomb in which he could have some confidence since they were prepared by the best scientific minds in Britain. Here were authoritative arguments for abandonment of plutonium production, thermal diffusion, the electromagnetic method and the centrifuge and calling for gaseous diffusion of U-235 on an industrial scale. Only in this way, said the report, could a bomb be produced in time to affect the outcome of the war. There was also the chilling warning that Germany had a head start and had been working on a bomb for three years. Roosevelt authorised Bush and the Uranium Committee to research the construction needs, the costs and the timeline to make a bomb and to return their findings directly to him but not to take any steps beyond research and development at this stage. With Roosevelt's blessing, Bush decided to hand over responsibility for the bomb project to the army even though they were less technically minded than the navy which would later claim that this decision 'slowed progress towards a bomb'.[11] Bush and Conant strengthening the S-1 with the addition of Fermi as head of theoretical studies and Urey as head of isotope separation and heavy water research.

Despite being officially excluded from the Manhattan Project, the navy still conducted its own research on nuclear reactors for possible use in ships and submarines at the Naval Research Laboratory (NRL) in Washington, D.C., under the leadership of Abelson specialising in the liquid thermal diffusion of isotope separation that had been discounted by Bush.

At the same time, Roosevelt realised that getting the British on board would be an essential step and a significant boost for the US programme. He exploited his close relationship with Churchill to suggest that they set up a joint Anglo-American effort to develop the atomic bomb despite the British having initially rebuffed US overtures in which they offered to pay for all future research and development, fearing that Britain would lose the technological advantage it held in early 1941. Hoping that the reality of Britain's penury and a growing realisation of the enormous resources that an atomic bomb project would consume, Roosevelt wrote to the British Prime Minister to try and change his mind. When Bush's two US delegates, Pegram and Urey, arrived in Britain in October 1941 they discovered that the British nuclear programme had been put under the Department of Scientific and Industrial Research and given the deliberately meaningless title of Tube Alloys Directorate. They went to Liverpool and visited Chadwick, who had written a report on 15 July 1941, focusing on the use of uranium as a means for producing a new type of bomb and a new source of energy. They were brought up to date on the latest developments and became convinced of the importance of the British research sharing their enthusiasm with Bush on their return. Whilst appreciative of the direction of British research they were less complimentary about British estimates of the bomb's power which seemed wildly optimistic at the same time as British estimates of the cost of producing one appeared to be on the low side. Bush then sent a report to Roosevelt on 27 November advising him that he was setting up engineering and scientific research groups to examine the practicalities of producing an atomic bomb. While Lawrence and Szilard were happy to see Oliphant stirring things up, Bush and Fermi still felt that the whole business was somewhat over-hyped and still feared being led into a costly and ultimately unproductive cul-de-sac. While Pegram and Urey's report was hitting Bush's desk, another account of their trip landed on Pavel Pastelnyak's. Pastelnyak (code name LUKA), was the resident Moscow hood, part of the Soviet *rezidentura* in New York. He had received this from an associate of Urey, the physicist and communist Emil Connison.

On 27 November Compton, echoing Churchill's sentiments, presented Roosevelt with the results of his review saying, 'the possibility must be seriously considered that within a few years the use of [atomic] bombs . . . may determine military superiority . . . Adequate care for our national defence seems to demand urgent development of this program.' It was not as optimistic as the Maud report but it strongly endorsed the US atomic bomb programme even if it did not bring results during the current war then as a tool of US diplomacy in the years beyond. Compton reported that two methods of isotope separation were thought to show promise, gaseous diffusion and centrifuge methods which would cost between $50–100 million but Bush had especially asked Compton to play down the importance of plutonium for fear of scaring off the politicians over the further enormous costs that line of research might lead to. Whilst the report showed the immense strides that research had taken in a very short time, there was still little to show in material terms. Only infinitesimal amounts of plutonium had been produced and no appreciable quantity of U-235 had yet been separated from U-238.

On 15 November German forces came to a frozen halt at the gates of Moscow and on 6 December the Soviets launched a massive counter-attack. On 7 December, Japanese carrier aircraft attacked the US naval base at Pearl Harbor and on the following day the United States declared war on Japan. On 11 December, Hitler, in a speech to the Reichstag, unilaterally declared war on the US. With war now a reality for the US the fear that Germany was ahead in nuclear research galvanised the NDRC. On 19 January 1942, a short note 'OK . . . FDR' written on White House notepaper was attached to Roosevelt's copy of the Compton report in what was probably the first official document signifying that a decision to build the bomb had been taken.[12]

Before this, Bush must have had a nod and a wink from the Oval Office to take preliminary steps. He had put Urey in charge of diffusion and centrifuge separation and heavy-water research, Lawrence took on electromagnetic separation and plutonium research while Compton ran fission chain reaction and weapon theory programmes. Abelson continued with his thermal diffusion research at the NRL. Bush retained responsibility for coordination of all engineering and construction contracts. He removed all uranium work from the NDRC and from this moment onward, broad policy decisions relating to uranium became the responsibility of a new Top Policy Group, composed of Bush, Conant, Vice President Wallace, Secretary of War Henry L. Stimson and Army Chief of Staff George Catlett Marshall. Following up on Roosevelt's initiative, Bush was directed to draft an official note to Churchill proposing that Britain and the US work together to develop a bomb based on the findings of the Maud report.[13]

This new impetus to atomic research brought the question of availability of raw materials roaring back into the limelight. At the top of the list was uranium ore without which nothing could be done. It was time to make a move to secure the Shinkolobwe output. US Office of War officials went to the Belgian Congo in 1942 and were dismayed to find Nazi influence on the rise with an increasingly anti-Semitic atmosphere and 'a lot of sentiment favouring doing business with Hitler'.[14] Straight away it was decided that the newly-formed Office of Strategic Services (OSS) under William 'Wild Bill' Donovan should try and put some level of permanent surveillance in place both to counter the German threat and find a way of securing the uranium supply chain to the US. Putting an agent into the Congo, however, was far from straightforward. It was a completely new environment for Donovan's men and required local knowledge. The man chosen to do the job showed just how thinly spread Donovan's resources were and what few options he had available to him. Armand Denis was a Belgian filmmaker who had worked in Hollywood before becoming a wildlife cameraman in Africa, India and Burma. Denis went to the Congo ostensibly to make a film about gorillas but his lack of training and dearth of support from the OSS resulted in his reports comprising little of economic or military use. He simply did not understand what he was supposed to be doing and was recalled to the US after only a few months, having provided no useful intelligence on Shinkolobwe. Denis would later become famous, alongside his second wife Michaela, as a maker of television shows about African wildlife. He was replaced as an OSS agent by an ornithologist, James Chapin, who operated under the OSS code name Crisp and was ordered to 'devise ways and means of expanding American intelligence operations to meet possible future developments on the African continent and to obtain military and economic information . . . that was not available through existing channels'.[15] That was quite an assignment for an ornithologist, albeit one with wide knowledge and experience of the Congo. While Denis had accepted his own ineptitude with equanimity, Chapin, who proved to be not only equally useless as a spy but a security risk to boot, found his role to be devastatingly stressful. He suffered a severe mental breakdown and spent months under psychiatric treatment in a Baltimore hospital.

Chapter 7

WERNER HEISENBERG

'If anybody in the world could make the atomic bomb, Heisenberg could.'
Ferenc M. Szasz, *'Peppermint and Alsos'*
No End Save Victory: Perspectives on World War II.

When Heisenberg had visited the USA in the summer of 1939, he had walked a fine line between enjoying the company of his friends abroad and treading the cautious path required to survive in a dictatorship at home. He was a proud German imbued with all the sensibilities of German scientific and nationalist traditions and culture who felt unable to turn his back on his country as it faced a future burdened with the consequences of years of catastrophically poor political decisions. He looked beyond that, however, to a time when it would be up to people like him to rebuild the nation's democratic foundations and in order to claim that right he was prepared to remain in Germany and endure the years of war from within the country. 'One day the Hitler regime will collapse', he had said to Goudsmit, 'and that is when people like myself will have to step in.'[1] Just how much he was ready to compromise with the Nazis in the meantime and create what he called 'islands of decency' has been the subject of much debate ever since. Paul Rose in his book *Heisenberg and the Nazi Atomic Bomb Project* asked the questions, 'did Heisenberg understand accurately how an atomic bomb would work?' and did he 'recognise the moral problem of becoming involved in the bomb project at all?'[2] Many, but not all, US and British commentators have taken the view that he generally ignored the moral issues surrounding his contribution to the Nazi regime's war effort.

In California he had seemed to Philip Morrison to be 'cheerfully oblivious of the crisis in Europe'[3] and maybe in the West Coast sunshine it was easier for him to ignore the dark clouds sweeping across the European continent but later, in Chicago, he was more open with Arthur Compton when he seemed to express a forlorn wish that the isolationist US might use its power to make Hitler think twice about starting a war that he had every expectation Germany would win. Having been prevailed upon by all the leading US and émigré scientists he met to abandon his country, Heisenberg had felt under a powerful obligation to go back.

Back in Germany on 26 September, he was conscripted into the War Office's Nuclear Physics Research Group under Paul Harteck to study the problem of building a 'uranium machine'. Heisenberg drafted a report in two parts under the heading *Die Möglichkeit der technischen Energiegewinnung aus der Uranspaltung* ('The possibility of the technical acquisition of energy from uranium fission') which was also known as the G-39 Report. The first part was submitted to the German Army Ordnance on 6 December 1939 and the other on 29 February 1940. Mark Walker in *Nazi Science: Myth, Truth and the German Atomic Bomb* says that 'After he had drafted his two-part G-39 report, Heisenberg no longer took an active immediate interest in uranium machines [and] left most of the actual theoretical and experimental research to [von Weizsäcker and] others'. However, Paul Rose claims that Heisenberg had wanted to build a nuclear explosive device as early as 1940 and goes to some length to show that the German army was thinking about a reactor bomb.

In their paper *The Theory of Nuclear Explosives That Heisenberg Did not Present to the German Military*, Carl H. Meyer and Günter Schwartz believe that Heisenberg had already concluded in his G-39 report that nuclear explosives could be built only with almost pure U-235, nearly impossible to produce in any quantity at the time, and fast neutrons and consequently doubted that Heisenberg ever seriously considered making a bomb. If research was ongoing on a reactor bomb, they contend that this was because Heisenberg gave his team a free hand to explore whatever lines they chose and could have allowed research into a reactor bomb even though he did not believe in it himself. It might also have been used as proof to keep higher authorities off his back by showing that he was exploring ideas that might be of great value militarily. It would have been fairly easy for a man like Heisenberg, if he so wished, to bamboozle his Nazi masters with language and concepts that would have gone right over their heads.

For Meyer and Schwartz, the G-39 paper clearly shows that Heisenberg was fully aware of how an atomic bomb could be built. He had all the knowledge necessary to develop the idea, but he either did not follow it up with the necessary theorising and calculation or, having done it, chose not to include the result in his reports. The fact that he did not provide such information, for whatever reason he may have had, prevented any serious research into building a nuclear explosive from the start even though there was a strong initial drive by a small group of pro-Nazi physicists, like Dieber and Manfred Baron von Ardenne, to get the project off the ground. The writers conclude that due to this lack of more precise technical guidance with respect to atomic bombs, Heisenberg, by this inaction, dampened any enthusiasm the German military might have had for developing an atomic bomb.

The purification of U-235 was a huge problem that seemed to defy solution, but work had started in July on a new building at the KWI to house an experimental nuclear reactor. Owing to its proximity to the Biology Department and the need to dissuade anyone from getting too close to it, the building was named the Virus House.[4] In late spring of 1940 in Germany, Harteck had used the Clusius-Dickel method to try and purify U-235 but would soon abandon it, as Frisch had done in England, having been unable to produce more than a miniscule amount of uranium hexafluoride, an essential intermediate chemical in the process. He had theorised that the limiting factor in the running of a uranium chain reaction was the level of the neutron multiplication achieved and it was he who came closest to creating such a chain reaction, but his experiment failed due to insufficient mass of U-235. Theoretically, if there had been a greater effort to acquire uranium oxide earlier, this quite likely would have led to the first ever chain reaction and could well have kick-started the German atomic bomb project. In a reference to an atomic bomb, Harteck concluded that separation of U-235 by Clusius-Dickel was feasible 'only for special applications in which [cost] is only a secondary consideration'.[5] Up to this point the Germans had pursued Clusius-Dickel with such blinkered concentration that it was only now that they began to look for alternatives but even so they still overlooked the gaseous diffusion method that Frisch was using in Birmingham. From now on the only option that seemed to give them any chance of progress was to use uranium oxide with enhanced but still low concentrations of U-235 and heavy water as a moderator.

The German atomic research programme was dealt two serious blows in 1940 with the deaths of Karl Becker, then president of the Reich Research Council, and Carl Bosch, president of the KWI, at a time when there was much discussion over how to organise science to support the German war effort. There was a lot of criticism of the KWI by those vying to take control of some of its institutions.[6] Heisenberg had objected to working under the Nazi Diebner, primarily because they had a mutual loathing but also because of Diebner's questionable credentials and for the way his 'clothes fit poorly, his glasses were always slipping off his nose and [for] his inveterate sycophancy'.[7] With the help of Wirtz and von Weizsäcker, Heisenberg had eventually managed to wrest from Diebner much of the day-to-day responsibilities of running the Uranverein which gave him the leeway, in early 1940, to instruct Paul Müller and Karl-Heinz Höcker to work under von Weizsäcker to get a better understanding of the critical mass issue and the use of heavy water as a moderator, graphite having by now been rejected as being much less efficient for that purpose as a result of Bothe's error. As a result, Müller presented a paper in May entitled 'A condition for the application of Uranium as an Explosive'[8] which clearly shows that the idea of a reactor bomb was still under consideration. In a subsequent paper a few months later he concluded that if a reactor is enclosed by a conventional explosive which is detonated then a chain reaction might result, but Heisenberg was later to say, somewhat disparagingly, that Müller's opinions were not taken too seriously in the KWI, despite his having co-authored a number of papers with him, sponsored and encouraged his work at the time and praised him for 'his valuable advice and discussion'. When looking at Heisenberg's post-war comments it is always worth comparing them with evidence from the time to test them for sincerity and veracity.

In March 1940 Harteck reported to the Uranverein that the two paths to making a bomb both seemed to be blocked. In the case of a bomb made of pure U-235, abandonment of the Clusius-Dickel method meant that the quantity required would be way beyond the capacity of industry to manufacture and as for the reactor bomb its sheer size and weight would make it totally impractical from a delivery point of view. It was just at this time, however, that von Weizsäcker brought new research to the table which was to have significant ramifications. He had been taking careful note of work done in the US especially that of Louis Turner on the discovery of a new transuranic element, plutonium, that had much more potential as an explosive.

At this point in the research an Austrian physicist, Friedrich Georg (Fritz) Houtermans, was co-opted into the German programme. Because of the free and open exchange of ideas and the frequent mixing of scientists through conventions and symposia, Houtermans had known Heisenberg and Fermi from his days studying at the University of Göttingen and Peierls had even been a witness at his wedding in 1931 but his communist politics got him into trouble when Hitler came to power in 1933 and he fled Germany, first to Britain and then to the Soviet Union where he worked in the Ukrainian Physics Institute in

Kharkov. When Stalin's purges started in 1937, he again found himself *persona non grata* but from a different perspective and was arrested and charged with being a German spy. After days of endless torture in the Lubyanka prison he signed a confession (it hardly mattered to what) to protect his family from persecution. He remained in prison for two years until he was handed over to the Gestapo as part of an amnesty brought about by the Molotov-Ribbentrop Pact of 1939 but when he returned to Berlin, having lost all his teeth in the Soviet prisons, he was imprisoned again on the Alexanderplatz. Pressure from fellow scientists saw him released in August 1940 when he went to work at the private Forschungslaboratorium für Elektronenphysik laboratory of the Nazi von Ardenne in Lichterfelde outside Berlin, but he remained under close surveillance by the Gestapo. Von Ardenne, who had only a rudimentary understanding of physics but was ever keen to take advantage of any government initiative and investment, had managed to get sponsorship to do nuclear fission research from the most unlikely source of the German Post Office. It was there that Houtermans began a research project on nuclear chain reactions based on the findings of McMillan and Turner that they had so injudiciously published a year before.

Throughout this research, Houtermans was in close contact with both Heisenberg and von Weizsäcker whose work underlay his paper 'On the Release of Nuclear Chain Reactions' and allowed him to conceive of a self-regulating reactor that was specifically designed to produce radioactive isotopes. Using liquid methane as a moderator, this reactor would operate at a low temperature and be an efficient producer of plutonium. It was clear to him that plutonium could be produced in a uranium fission reaction and could then be easily separated out and used as the basis for a bomb. What he did not include was any mention of uranium critical mass or a recognition that a bomb would require fast neutrons. While there is evidence that Heisenberg understood that the rate of neutron production was critical, Houtermans had made no attempt to calculate it. The fact that von Ardenne was not a trained physicist made it easier for Houtermans to play down the significance in his reports of any aspects of his work that he did not want highlighted. Von Ardenne was never likely to work out for himself the significance of plutonium as a route to an atomic bomb.

Houtermans had a meeting with von Weizsäcker and Heisenberg in Berlin in early 1941. Heisenberg told Houtermans that it was his ambition to 'guide German science safely through the war' which did not include the building of an atomic bomb.[9] Houtermans confided that, because of his precarious security status, he felt trapped into doing work on chain reactions that had military applications. Whatever the nature of further discussions between the three men it cannot be doubted that after their meeting all discussion of plutonium became very muted and research along the lines of a plutonium bomb was supressed from within the German scientific community.

When von Ardenne persuaded Heisenberg and others to visit his laboratories, Houtermans took the opportunity to voice his reservations to him and his shadow, von Weizsäcker, over the morality of going further down the road towards an atomic bomb. Heisenberg and von Weizsäcker would later claim, when defending themselves against accusations of helping the Nazis to build an atomic bomb, that they had never been close to von Ardenne, a man who, they said, 'would produce a bomb for anyone'.[10] Von Ardenne, however, asserted that it was Heisenberg who, knowing quite well that von Ardenne was working with the Nazis, gave him the crucial piece of information from KWI that it was possible to make an atomic bomb using only a few kilograms of fissile plutonium.

Houtermans was not at all comfortable at the way his work was being used as part of the German atomic bomb project and even tried to send a message to the US scientific fraternity by way of Fritz Reiche who had been given permission to leave Germany, after years of trying, to emigrate to New York in March 1941. The 'quiet and gentle' Reiche was a brilliant theoretician and close friend of Einstein but had been forced into premature retirement at the age of 50 as a consequence of the 1933 Nazi racial laws. He still met Hahn, Heisenberg and others occasionally but was completely debarred from holding any academic position. His son was allowed to leave Germany before the war but his daughter had become too ill to travel. After laboriously acquiring the many documents required, he was granted consent to emigrate just weeks before all emigration of Jews was banned in Germany. When Houtermans visited him on the night before his departure his message was short and blunt. It clearly contained details of military secrets that would have ensured a quick end for him, Reiche and Reiche's family had it been intercepted by German security. Heisenberg was working on a bomb but trying to slow down the research, he told Reiche, and the US should accelerate any similar programmes they were pursuing.[11]

Reiche and his family made the 36-hour rail trip to Lisbon where they boarded the SS *Excalibur* which took them along with other Jewish refugees to Ellis Island in New York. While he was assiduous in passing Houtermans' message on to a number of eminent scientists include Szilard, there is no record of any of them ever bringing it to the attention of anyone in the higher reaches of the US uranium project.

This attempt by Houtermans to alert Western leaders to the Nazi atomic bomb menace may have been well meant at the time but his actions subsequent to the German invasion of the Soviet Union when he followed behind and systematically looted Soviet laboratories suggest that his antipathy to the Nazi cause was disappearing. Rosbaud, who had counted Houtermans as a friend, would later find him 'morally compromised' and his behaviour 'disgusting'.[12] Despite his earlier communist sympathies, Houtermans' treatment in the Moscow cellars had clearly stripped him of any lingering affection for the Soviets. His later wartime actions can be seen more in that light rather than in any new pro-Nazi affiliation.

Both Heisenberg and von Weizsäcker kept in close contact with Bohr in Copenhagen. Bohr had continued to work there and since Denmark had been allowed more freedom than other occupied countries, he had been able to maintain informal contacts with scientists in Britain and the US but was resisting pressure to collaborate with the German nuclear programme. In the spring of 1941, von Weizsäcker visited him to discuss progress on the nuclear front but colleagues of Bohr were reticent to engage with him, believing that he had been sent to spy on them. When he returned to Germany, von Weizsäcker seemed to confirm this by reporting to the army that Bohr was definitely not working on uranium fission but, in the US, Fermi definitely was. Bohr, he said, appeared to be unaware of the direction of nuclear research in Germany and von Weizsäcker had not enlightened him.

In September 1941, it was Heisenberg who travelled to Copenhagen ostensibly to participate in a German-sponsored scientific conference at the Institute for Theoretical Physics, but it was really to meet Bohr who habitually refused to attend any joint German-Danish scientific conferences, believing them to be no more than Nazi propaganda. Bohr, however, did agree to meet with Heisenberg during his stay although right from the start he suspected that it was little more than a 'fishing trip' to find out what he knew about research and development in Britain and the US and even an attempt to inveigle Bohr into some sort of collaboration with the German programme. Bohr recalled that Heisenberg told him of his and von Weizsäcker's definite conviction that Germany, after its lightning victories in Poland and France and now smashing its way to the gates of Moscow, would win the war, which would be quite a good thing and that it was therefore quite foolish for anyone to maintain the hope of a different outcome. Bohr had admired Heisenberg for his efforts to help Jewish scientists in the 1930s but was shocked when the same man now refused to apologise for Germany's actions in Poland by saying that at least they had not been so brutal in

France. Bohr also got the distinct impression that under Heisenberg's leadership, everything was being done in Germany to develop atomic weapons and that for two years he had been working more or less exclusively on such preparations.[13]

Despite what Bohr said, Heisenberg claimed that he told him that it was not acceptable to put an atomic bomb in Hitler's hands but his dilemma was what to do about it. His solution was to continue working on the atomic bomb project and, by retaining his dominant status as Germany's foremost scientist, to exert influence over the direction it was taking and keep it, as far as possible, out of the hands of the military.[14] It was just before the Bohr meeting that Heisenberg claimed that he had fully realised that plutonium was no longer just a theoretical novelty and, if given prominence in German research, would create what he called 'an open road to the bomb'.

Heisenberg would later claim that his primary motivation for meeting Bohr, who was something of a father-figure in the scientific world, was to discuss the whole subject of the morality of developing an atomic bomb, implying that Heisenberg had reservations and that if Bohr was strongly against it then it would encourage him to scale back his own involvement. Heisenberg would later say that his central motive in bringing up the atomic bomb in discussions was to indicate to Bohr in a roundabout way that Germany had not made any great effort to develop its own bomb and that Bohr could use his contacts to assure the US that they had nothing to fear in that regard, hoping that might discourage any attempt to pursue its own atomic research to build its bomb. He clearly feared that a bomb could now be built, that the US had the resources to do and that it might use it against Germany. A more optimistic outcome might be to look beyond the war to a nuclear-free *Pax Germania* when international understanding would once again be established amongst physicists and Germany would be re-established in its place at the head of the table. Had he been a little more worldly he would have realised that the US was also thinking about the post-war world and once it set out on the nuclear path nothing would prevent it reaching the end. When the conference opened, a number of scientists were shocked and dismayed to hear Heisenberg express the view that for Germany to win the war would be no bad thing if it scaled back the threat of communism and brought peace and security to Germany's eastern borders.

Bohr's wife, Margrethe, clearly recalls that Heisenberg's visit had been 'hostile' and certainly the warm friendship that the two men had enjoyed before the war seems to have been severely tested by the exchange. Subsequent relations were said to be civil but strained.[15]

Bohr's recollection of the meeting was given in a letter he wrote to Heisenberg in 1958 in which he said:

> Personally, I remember every word of our conversations, which took place on a background of extreme sorrow and tension for us here in Denmark . . . If anything in my behaviour could be interpreted as shock, it did not derive from such reports but rather from the news, as I had to understand it, that Germany was participating vigorously in a race to be the first with atomic weapons . . . at that time and ever since, I have always had the definite impression that you and Weizsäcker had arranged the symposium at the German Institute, in which I did not take part myself as a matter of principle and the visit to us in order to assure yourselves that we suffered no harm and to try in every way to help us in our dangerous situation.[16]

The meeting with Bohr is significant because it came at a time when German research was hitting a barrier, with the Nazi regime reluctant to invest the kind of finance and resources in the atomic bomb project that would allow it to move forward. Heisenberg would have been acutely aware of that and he would also have been under no illusion about the kind of support that might be available for development of a bomb in the US if the government there was taking it seriously. If he was sure that a bomb was a practical possibility that might become a reality before the end of the war, as he implied to Bohr, he had a very good idea which side would have it.

Heisenberg's version of the meeting had him, at first, trying to assure Bohr that German victories against the Soviet Union would benefit the whole of Europe and then expressing a wish that Bohr would return with him to Germany where his friends were better placed to protect him, without appearing to understand what an unthinkable thing it would be for Bohr to do. At least in Copenhagen he had some semblance of independence and a choice of the sort of research he was undertaking, which would not be the case within the borders of Nazi Germany. Broaching the subject of an atomic bomb, Heisenberg asked Bohr's opinion on the morality of building atomic weapons. Bohr took this to mean that Germany was doing just that and Heisenberg was looking for some sort of justification.

Bohr felt that he was being interrogated about his knowledge of developments in Britain and the US and began to wonder if Heisenberg was working for an authority outside the scientific community. Relations between the two men became strained. Bohr was indignant that he should be interrogated in such a crude fashion and must have showed it as might be gauged from a remark that Heisenberg made to a colleague days after the meeting when he admitted that Bohr had 'not understood anything' that he had said and that the conversation had gone 'badly wrong'.[17]

Heisenberg's failure to convince Bohr threatened to have severe consequences. He had clearly let slip to Bohr that atomic research was ongoing in Germany and in so doing had exposed the serious risk that this information would find its way to Britain and the US via Bohr's unofficial contacts there. If it could ever be shown that Heisenberg had been even peripherally complicit in this security breach it would have had a fatal outcome for him, his family and a great many of his colleagues. The breakdown of relations with Bohr affected Heisenberg badly and his inability to persuade Bohr to join in a common cause against all development of a bomb weighed heavily on him. Heisenberg's friends thought that the reason the meeting had gone so badly was that Heisenberg had been too much in awe of Bohr to speak clearly and freely, which left much of what passed between them open to ambiguous interpretation. Far from presenting Bohr with a precise idea of his motivation and objectives, Heisenberg had allowed Bohr to interpret what he had said in whatever way he chose and Bohr had chosen to see the worst of it. Plans were laid to make a fresh approach but events in Russia changed the picture. When the German assault on Moscow was halted and the German armies in the Soviet Union faced a bitter winter out in in the open, the mood of triumphalism in Germany was badly shaken. An early victory in the East was now out of the question and Hitler retrenched for a long war.

In late 1941, von Weizsäcker had applied for a patent, the details of which were later recovered from Soviet archives, that shows he clearly understood the potential of plutonium as an explosive and its advantages over U-235 so Heisenberg would have been fully aware of them also. Hans Bethe would later say that von Weizsäcker's patent was for nothing less than a plutonium bomb. The question therefore is why the German military did not pursue investigations into making one but a bigger question arises when details of a second patent submitted by von Weizsäcker, but this time in the name of the KWI, contained no mention at all of a nuclear bomb.

Pressure was building on the German Army Weapons Research Office to come up with evidence of progress towards any sort of super-bomb, whether atomic or the smaller reactor bomb, with a threat to hand over control of the nuclear programme to the RFR. All the leading nuclear scientists were called together on 16 December and asked to estimate the amount of time required to come up with a demonstration of practical results. The impact on the German economy reached the Uranverein in the form of a stark warning from *Heereswaffenamt* director of research, Erich Schumann, to the effect that its work could only continue if 'there is a certainty of getting some benefit from it in the near future'.[18] This was the critical point of the German atomic bomb programme. Schumann, according to Heisenberg, was a time-serving careerist who had 'no firm grasp of fission or its implications' and left all scientific matters in the hands of Diebner whose penchant had always been for a military application of atomic power.[19] The military representative, General Emil Leeb, quickly concluded that the army should wash its hands of atomic research. There would be a six-month evaluation during which at least six different agencies would vie for control of the atomic programme before it came within the purview of the Minister of Armaments and War Production, Albert Speer. The essential question that was pored over was simply whether or not atomic research was worth investing in at all.

Schumann asked Diebner, as head of the Nuclear Physics Section of the Weapons Research Office, to carry out a complete survey of the project. Diebner's team came back with an aggressively optimistic 144-page report in February 1942 that, unsurprisingly, recommended that the programme be elevated into a full-scale industrial project despite acknowledging that U-235 separation on an industrial scale or the manufacture of plutonium were seen as very long-term projects that were virtually impossible to realise before the war would end and, oh yes, not to mention the critical shortage of raw materials.

Control of uranium research was handed over to the Education Ministry, freeing it from military control even before the HWA report was submitted, and the new body soon organised a conference to provide a showcase for nuclear power. It took place at the Harnack-Haus of the KWI Society for the Advancement of Science (Kaiser-Wilhelm-Gesellschaft zur Förderung der Wissenschaften, KWG) between 26 and 28 February. Heisenberg delivered a lecture entitled 'The Theoretical Foundation for the Production of Energy from Uranium Fission' and the German Army simultaneously published a report entitled 'Energy production from uranium'. Even though discussion of a bomb was not precluded, together these suggest that the main focus of research at this time was not on an atomic bomb but on generating electricity or propulsion mechanisms. Rose believes that Heisenberg's presentation contained some fundamental misconceptions about the physics of an atomic bomb and that he had clearly misunderstood, or deliberately misrepresented, the issue of how big a critical mass would be required to make a bomb but his estimates differed little from those prevalent in the Smyth report in the US at the same time so not too much should be made of that. He did not ignore plutonium as a fissile material and touched on its importance when he described it to the audience of politicians and military leaders as 'being an explosive of equally unimaginable power as pure U-235' but said little about how the plutonium could be produced. The essence of his approach to plutonium seems to have been that it was very interesting but really of no practical significance in the context of the current war. His presentation might have had more effect had many senior military commanders chosen not to attend having been given, deliberately or otherwise, the wrong programme that had promised not a breakdown of current atomic research and its application but an exceedingly boring-sounding 'twenty-five highly technical scientific papers'.[20]

Heisenberg's recollections of the meeting were that he told his audience that an atomic bomb was entirely possible but that it could not be built within the anticipated duration of the war even if he had under his control 'the best researchers . . . and a large part of the industrial resources of the country'.[21] In the light of the military setbacks in the Soviet Union that meant a severe curtailment of investment in all but those projects that were of immediate benefit to the war effort, his apparent *volte face* in his February 1942 presentation as opposed to the line he took with Bohr may well have been because he now began to think in terms of a German defeat and may have considered that any involvement he might have with an atomic bomb programme might not go down too well when all were brought to account. Whatever his motivation, it seems clear that from this point there were two competing groups working on nuclear reactors; Heisenberg's and Diebner's.[22]

On 4 June, Speer gave Heisenberg the perfect opportunity to state his case for atomic power before a distinguished audience in the Helmholz lecture room of the Harnack-Haus. Speer's interest had been aroused during a conversation with General Fromm who told him of a weapon that could 'annihilate whole cities'[23] and he was sufficiently impressed to invite top civil servants and military advisers to join Fromm, Leeb and Schumann from the *Heereswaffenamt*, Milch from the Luftwaffe, Rhein and Witzel from the Kriegsmarine along with all the leading atomic scientists in the country. Presumably it was Speer's reputation and authority that had persuaded so many leading figures to attend because few outside the scientific community were aware of research into atomic power and had little idea of what to expect from the meeting.

No first-hand record exists of the meeting, only recollections recorded years later, but it seems that Heisenberg delivered a rather bland lecture on nuclear power from a purely scientific perspective which would have gone right over the heads of most people present. Given the high-ranking military presence, he might have been expected to drive straight into the military potential of atomic power. If he really wanted to take full advantage and boost funding for his research and investment in the industrial infrastructure required to build a bomb, these were the people he needed to impress. Instead, Speer recalls Heisenberg spending time berating the military for drafting so many young scientists into the fighting war and bemoaning the lack of funding for scientific research in general rather than trying to generate enthusiasm by describing the destructive power of a plutonium bomb. Speer must have been disappointed with the way the meeting had seemed to drift away from the main issue he had hoped would be addressed. Many in the audience must have begun to wonder why they had ever been invited. Many would have wished that they hadn't bothered to accept. It was left to a clearly frustrated Speer to try and steer the meeting round to discussing a bomb and when he asked how atomic science could be applied to the manufacture of bombs the audience was visibly and audibly surprised at the very notion.

They were much more attentive now as Heisenberg assured them that 'It does not appear impossible that we will one day [have] explosives that are more than a million times more effective than those currently available'.[24] He did, however, warn that realisation of such a thing would be 'enormously expensive and . . . would take perhaps many years'.[25] In reply to a question by Milch who wanted to know how large a bomb would have to be to destroy a large city, such as London, Heisenberg, somewhat misleadingly is reputed to have said that it would be only 'about the size of a pineapple' which was obviously a reference to the critical mass and not the whole bomb. This was a staggering assertion that stunned the audience into silence but Heisenberg can hardly have had confidence in such a claim and may well have been guilty of 'playing to his audience'. If he had overstepped the mark there, he was quick to row back and disabuse his audience of any notion that a war-winning device was just around the corner but also made it clear that he believed that the Allies had come to similar conclusions and were equally far away from a having an atomic bomb.

Later, during dinner, Heisenberg took the opportunity to ask some of the military leaders how they thought the war was going. All gave bland assurances but none were convincing. Even Speer avoided giving a straight answer. Speer's bitter disappointment at the way the meeting had gone, however, did not prevent him from offering increased financial aid to the atomic project and he was the one man who could make a significant difference. He raised the bomb issue with Hitler for whom, Speer says, the idea 'obviously strained his intellectual capacity' since he was 'unable to grasp the revolutionary nature of nuclear physics'.[26] Speer could have transformed the bomb programme in an instant by allocating substantial resources to it as he would do to the V1 flying bomb and V2 rocket programmes. When Heisenberg came back with a 'ridiculously tiny'[27] request it left Speer confused, having heard him tell the 6 June meeting how vastly expensive it would be to develop atomic power on an industrial scale. He had been the last hope that Germany would seriously compete in the atomic race and his loss of faith in the programme was evident by arranging, with Hitler's approval, to transfer control of the RFR to Hermann Göring, who had no understanding of or interest in it. The modest funding that Speer had facilitated had given Heisenberg everything he had asked for but for Speer it was the end of the road.

Heisenberg now concentrated on designing a natural uranium reactor for the production of energy and also as a doorway to plutonium research but he claimed that after his conversations with Speer, he and other German scientists no longer had to grapple with the dilemma of working towards an atomic bomb for the Nazis. Uranium enrichment was not entirely abandoned but was not pursued with any great expectation of success. Research in other areas was split between a number of institutions and work was directed according to the agendas of each, essentially putting it beyond central control.

Although the idea of a bomb had not been totally discounted, the main thrust of research in Germany was now geared towards building a working reactor to generate power for industry. Three test reactors had failed and a fourth, L-IV, was set up in Leipzig but was delayed until enough heavy water was brought down from Vemork. When research resumed, catastrophe struck. An 'atomic pile' of five tonnes of heavy water and 10 tonnes of uranium oxide in a spherical chamber was set up to see if it could sustain a fission reaction. The 'pile' was immersed in water in which, after 20 days, a small stream of bubbles emerged indicating that the reactor had a small leak. On 23 June 1942 the spherical reactor was lifted clear of the water for inspection. A technician opened one of the inlet valves and allowed air into the chamber, causing the uranium powder to burst into flames giving out intense heat. The chamber was immediately re-sealed and lowered back into the water bath and the fire appeared to have been extinguished. Heisenberg was called and confirmed that the situation was now under control and he left, but the situation was far from under control. He had to come back soon afterwards when the reactor chamber had started shuddering and swelling up. It was obvious what was going to happen and all the staff just managed to escape the room before intense heat caused the cooling water in the jacket to flash into steam and the reactor chamber exploded. Particles of the uranium heated to a temperature of 1,000° Centigrade shot up 6m to the ceiling and showering across the laboratory building, setting it on fire. There had occurred the first nuclear accident in history. While the experiment had gone wildly out of control it at least confirmed the explosive potential of uranium.

A further meeting was held on 6 July 1942 to discuss the function of the RFR and set its agenda. Copies of a speech delivered to the meeting that were uncovered in KGB archives in Moscow indicate that at this time Heisenberg was strongly in favour of continuing with research into nuclear weapons, perhaps because he knew that it was now impossible. He probably tried to appear enthusiastic both to ensure that he would not be dragooned into alternative war work and also probably to satisfy his own scientific curiosity. He set about getting a uranium reactor working in the hope of producing plutonium. His methodical approach meant that only slowly were a number of trial reactors set up using uranium and heavy water to try and create a chain reaction. The project thereafter maintained its *kriegswichtig* (important for the war) designation and funding continued from the military, but it was now broken down into a number of different institutions and put under the control of the 'scientifically illiterate' SS Obergruppenführer Bernhard Rust, the very man who had blocked Lise Meitner's exit visa in 1938. The high point of the German nuclear energy project had been in early 1942 when measured by the number of people devoted to the effort. After this, the number of scientists working on applied nuclear fission diminished dramatically and many of the scientists not working with the main institutes were restricted to devoting their efforts to more pressing war-related work.

Heisenberg now set about working on a reactor to act as an engine for power production but the idea of a reactor bomb was always lurking in the shadows. One month after the 6 July meeting he wrote a paper outlining a proposed large-scale reactor experiment (G-161) in which 'the whole fission disintegration of the uranium nuclei in an explosive fashion will proceed in a fraction of a second'.[28] When the Nazi physicist Rudolf Fleischmann petitioned for funds to build a powerful new cyclotron he was careful to point out that not only would a uranium machine make an excellent engine but 'by means of a rapid ignition [would take on] the character of a bomb'. Flügge had also refined his theory of a bomb and concluded that 'a considerable reduction in the size of the plant . . . greatly extends the scope of application in a Uranbombe'.

When attempting to evaluate Heisenberg's attitude to making a German atomic bomb evidence is scarce and often contradictory. One series of reports that throws light on the issue comes from British Intelligence archives in relation to messages coming out of Germany in late 1942. These were from disaffected German scientists working on the atomic programme who took enormous risks by discussing it with people not only outside the scientific community but outside the country. There was hardly any sort of British intelligence network inside Germany at the time; the only Allied spies were working for the Soviet 'Red Orchestra' network which had no contacts inside the Uranverein. While the British could not get into Germany, however, German scientists could get out and a few made trips to research establishments in Scandinavia such as Bohr's laboratories while others kept up contact with the heavy water plant at Vemork where new methods were constantly being tried out to boost production. The neutral Swedish capital of Stockholm became a hub through which many of the visitors to Scandinavia passed. Here also were Allied agents and the headquarters of resistance groups operating inside Norway and Denmark.

Einar Skinnarland was a young Norwegian engineer working on the Mos Lake dam that supplied power to the Vemork fertiliser plant and had extensive contacts in the plant. He had been trained in espionage by British Intelligence and formed, with the engineer Jomar Brun, an intelligence cell inside the Vemork plant. Brun had extensive contact with German scientists such as Karl Wirtz and Hans Suess who had discussed with him the German atomic bomb programme and the crucial part that Vemork played in it. Suess assured Brun that the rate of progress towards a German bomb was putting it well beyond the probable duration of the war. He was playing a dangerous game, knowing full well that all this was being relayed to London by Skinnarland. Harald Wergeland was a friend of Wirtz and also a member of the Norwegian underground along with Njal Hole, a scientist working at the Manne Siegbahn laboratory in Stockholm where he was in daily contact with Lise Meitner.

Another Swedish physicist who had occasional contact with Lise Meitner and Niels Bohr wrote to a friend in London, who duly passed it on to British Intelligence, that he understood German scientists to be working under Heisenberg to create a chain reaction using U-235. A third source was Hans Jensen, a colleague of Paul Harteck, who had been urged to follow up on Heisenberg's disastrous meeting with Bohr by making a second trip to try and repair the damage done. He went to Copenhagen but Bohr, acting on advice from British Intelligence agents that Jensen was most likely an *agent provocateur*, did not trust him and would not see him at first. However Jensen did meet Christian Møller and told him all he knew about work being done in Germany with heavy water and was in fact on his way to Vemork to work on ways of increasing supplies. When Bohr later changed his mind and had a meeting with Jensen he told him how angry Heisenberg had made him at their meeting in September. Jensen then continued to Vemork where he met members of the Norwegian underground and told them that there was very little progress in Germany towards an atomic bomb, but research was ongoing with regard to atomic power. Heisenberg, he said, was sure that Germany would not make a bomb. Wirtz, who was said to be close to Heisenberg, went to Norway soon after the Jensen visit and reiterated the message that 'the goal of research [in Germany] was a power-producing machine, not a bomb'.[29] The Norwegian Resistance offered to spirit Wirtz away to England but he declined.

While all this was giving a clear, unequivocal message that the Allies did not have to fear a German atomic bomb, a new contradictory report came in from the US where Szilard had picked up a rumour emanating from a contact he had in Switzerland and had passed it on to Arthur Compton who in turn passed it on to Tube Alloys in London. The disturbing story was of the Germans having 'got a chain reaction going' and this was a good six months before such a breakthrough would be achieved by Fermi in Chicago. Compton's warning of a 'real danger of bombardment by the Germans within the next few months using bombs designed to spread radioactive material in lethal quantities' had now passed through a number of links but was greeted with great alarm and given much attention.[30] The British, replying to Conant poured cold water on the intelligence coming out of Scandinavia in favour of the Szilard rumour and said 'we don't regard [statements about Heisenberg's reticence] as of much significance'. Apparently, it was thought much safer to believe bad news than to put too much faith in good news.

Chapter 8

THE MANHATTAN PROJECT

'It would take the entire effort of a country to make a bomb.'
Enrico Fermi, 1939

Back in Britain, Oliphant and other leading Maud scientists were dismayed to receive notice that the government had handed over the running of the Tube Alloys Committee to Wallace Akers, the head of an ICI team that had previously unsuccessfully lobbied to take over the whole Tube Alloys programme. Oliphant, believing that it should have been he who got the top job and angry that the project was now run by someone 'completely ignorant of the essential nuclear physics upon which this whole thing is based', complained that Akers would look after the commercial interests of ICI before the needs of the country and resigned on the spot. It was a relief for him to get out from under what he called the busybodies of committee politics.[1] Chadwick, on the other hand, found the new arrangement quite acceptable and persuaded Oliphant to give it time which he did and eventually mellowed after lunching with Akers, in the best English tradition, and finding him after all to be 'an excellent person to be in charge'.

Despite promising results from plutonium work at the Cavendish, Britain was forced to concentrate its limited resources on the purification of U-235. In the US, however, by early 1942 plutonium research at Berkeley had started to show promise and was given equal attention to the three pathways towards uranium purification – gaseous diffusion, electromagnetic separation and centrifugal separation. By May 1942, there was no obvious leader in the race to provide the crucial fissile material for a bomb.

A plutonium bomb might provide a quicker route, but it was not yet clear that plutonium could actually be produced in a uranium pile and then be separated in usable quantities. Compton tried to clarify this at the new Metallurgical Laboratory (Met Lab) at the University of Chicago by conducting a review in May 1942. By now security was of the essence and to ensure secrecy, the whole nuclear programme was handed over to the Army Corps of Engineers. In the summer of 1942, Colonel James Creel Marshall was put in charge of the Laboratory for the Development of Substitute Metals (DSM), the military's initial cover name for the project. Marshall changed that to the Manhattan Engineer District (MED) on 13 August.

Bush and the Army chiefs agreed to appoint Colonel Leslie R. Groves to oversee what was now referred to as the Manhattan Project on 17 September 1942. Six days later, at his own insistence, he was promoted to brigadier general, giving him the rank he felt he needed to get the job done. Groves, whose impressive credentials included overseeing all army construction in the United States, a mammoth task involving building camps, munitions plants, airfields, depots and the Pentagon, moved with commendable speed and relocated the MED headquarters from New York to Washington. Having concluded that the whole project needed to be better coordinated, Groves ordered the setting up of production plants for U-235 at Oak Ridge (Site X) in Tennessee and, despite concerns about his administrative inexperience and left-wing politics, appointed J. Robert Oppenheimer to lead the research and design laboratory at Los Alamos in New Mexico (Site Y) where theoretical and experimental work could be conducted according to standard scientific protocols. Given how closely these two men were going to have to work it was a concern that they could hardly have been more different, but their professionalism and maturity ensured that their relationship was underlain by a good deal of mutual respect.

A number of locations for the nuclear site had been considered over the summer, including Shasta Dam in California, two spots outside Chicago and several areas in Washington State, as well as a few other locations in East Tennessee. However, all of these potential sites were deemed either too small or too vulnerable to enemy attack. Oak Ridge was considered suitable because of its rural location in a landlocked state. East Tennessee had just the right size of population for the Manhattan Project's purposes. It was not too densely populated but had enough people to work at the new plants. The plan was to build four different types of production plants on a single site which demanded significant amounts of electricity. This could be provided by Tennessee Valley Authority hydroelectric plants at Norris Dam. As an added bonus, the land was cheap: only $48 per acre, on average.

Under orders from General Groves, the US Army Corps of Engineers purchased 59,000 acres of land along the Clinch River. Originally known as Site X, the nuclear site was eventually renamed Oak Ridge. With its promise of jobs, the new city drew in tens of thousands of families, becoming the fifth largest city in Tennessee. The entire town and facility were fenced in, with armed guards posted at all entrances. Workers were sworn to secrecy and only informed of the specific tasks they needed to perform. Most were unaware of the exact nature of their final product. The engineering at Oak Ridge would be made extremely difficult by high levels of radiation produced by the process. Radiation safety was a major consideration because of the hazards of working with plutonium, which was highly radioactive and lethal if ingested. Uranium, a much less radioactive element than plutonium, posed far fewer safety problems. Oppenheimer suggested that the bomb design laboratory should operate secretly at a secluded location and Groves began looking for an appropriate site. By the end of the year, he settled on the Los Alamos Ranch School, an isolated boys' school on a mesa high in the Jemez Mountains.

His credentials proved that Groves was a self-confident, capable and practical man, but on a personal level, he was brusque, ruthless and egotistical. With abrasive and sarcastic directness, he drove his team relentlessly without the niceties of regular praise to uphold morale. His aide Colonel Nichols called him 'the biggest son-of-a-bitch [he] had ever worked for'. Oppenheimer on the other hand was a philosophical man with a sensitive nature, a chain-smoker given to working long hours. He was concerned to look after the emotional and intellectual needs of his highly distinguished staff. Once described as a 'cocktail-party bon vivant who prided himself on his skill at making perfect martinis, who possessed a penetrating intellect and who took pains to be eloquent regardless of the subject, be it quantum physics, Sanskrit religious texts, seventeenth-century French poetry or some point in the third volume of Karl Marx's *Das Kapital*', he was intolerant of those who could not keep up with his rapid-fire analysis of subtle and complex physics problems and puzzles. Groves, on the other hand, was a 'bluff and blustery teetotal engineer who knew how to get things done and who had no time for idle speculation about the nature of the universe or the nature of truth, beauty and justice, a man instinctively distrustful of any political opinions to the left of Herbert Hoover'.[2]

Groves had a justified reputation as a man who got results and for having an unerring talent for choosing the right people for the job. He would travel halfway across the continent to intervene personally, and where necessary brutally, when any of his projects threatened to stall. 'I do not want the sun to set tomorrow with this man still in Washington', was one of his more memorable orders regarding a particular subordinate who failed to live up to expectations.[3]

When Groves took up his post, research on plutonium had barely begun. He was not encouraged by being told that there was as yet no experimental proof that it could actually be produced on an industrial scale. Nevertheless, he forged ahead with plans for a major production site and called a meeting of all the top atomic scientists at the University of Chicago. Groves later wrote that, at the time, he was most concerned with the gaps in the scientific knowledge so that he got a feel for the scale of the task ahead of him. The whole of the building programme would be determined by the amount of plutonium and U-235 that was required. As a man given to tackling practical problems with a fair idea of where he wanted to end up, Groves was astounded to be told that a bomb would require anywhere between ten and a thousand pounds of fissile material, a range which, in his own words 'completely destroyed any thought of reasonable planning'.[4]

Whether it was at this meeting or soon afterwards, Groves, for some reason, took a particular dislike to Szilard who, afterwards, referring to Groves, asked a colleague 'How can you work with people like that?'[5] Clearly Szilard's well-known disdain for authority and general prickly character had quickly manifested itself to Groves and irked him. He initiated an intensive investigation of Szilard's background and concluded that he was untrustworthy and uncooperative, 'the kind of man that any employer would have fired as a troublemaker'.[6] According to records of the Army Corps of Engineers, Record Group 77, National Archives and Records Administration, College Park, Maryland, it got to the point where Groves drafted a letter to the US Attorney general saying:

> The United States will be forced without delay to dispense with the services of Leo Szilard of Chicago, who is working on one of the most secret War Department projects.
>
> It is considered essential to the prosecution of the war that Mr. Szilard, who is an enemy alien, be interned for the duration of the war.
>
> It is requested that an order of internment be issued against Mr. Szilard and that he be apprehended and turned over to representatives of this department for internment.

Either the letter was never sent or its recommendations were ignored because Szilard continued to be part of the Manhattan Project although one can be certain that Groves would ensure that he was very carefully monitored by the security apparatus. 'The investigation of Szilard should be continued despite the barrenness of the results,' he said, having found no actual evidence that Szilard was a security threat.

At the time of Groves' appointment in the summer of 1942, there was serious concern about the German atomic bomb programme. There was a widespread belief among US scientists that the Germans had a significant head start on them. As early as 1939 the evidence of German activity brought to the US by scientists returning from Europe pointed to a coordinated and intensive research programme, centred on the KWI in Berlin with special focus on the separation of uranium isotopes. Then Flügge published his article in *Die Naturwissenschaften* with a more accessible version printed in the German newspaper *Deutsche Allgemeine Zeitung*. Together, these articles clearly showed that there was great interest in an atomic bomb within the German scientific community. There were also a number of European scientists who had remained in Europe during the war who, although opposed the Nazi regime and not directly working for them, were privy to much of what was going on at the KWI and covertly passed along information about German scientists to their American, British or émigré colleagues such as Szilard.

Germany now also had many of the necessary facilities, materials and industrial infrastructure needed to start up a successful atomic bomb project. German industry had a formidable reputation in the US and few doubted that if they dedicated a significant proportion of their industrial resources to an atomic bomb project it would be successful. Germany also had many of the world's greatest scientific laboratories, including the KWI and laboratories at the Universities of Gottingen, Leipzig, Cologne, Hamburg, Giessen, Heidelberg and Vienna, not to mention Bohr's institute in Copenhagen and the Joliot-Curie laboratory at the Collège de France. The US fully appreciated the value of heavy water to any nuclear programme but had avoided going down that route because of its rarity. The Germans, however, had access to the only operational heavy water plant in Europe at Vemork. Bush was in no doubt that it would be very difficult to obtain information about enemy activity and that any attempt to estimate when a German bomb would be available would be purely speculative. All they could do was put together scraps and fragments of information that might be useful in building up a bigger picture.

Because of the tremendous military potential of atomic research, it was accepted right from the beginning that there would be a requirement for maintaining a high degree of secrecy and it was this which had prompted the government to hand over administration of the programme to the Army which, they felt, was best placed to enforce security. With that decision a whole separate secret state within a state was created. It was to be a mammoth task to maintain secrecy given the numerous contracts with industrial firms, the employment and interaction of many thousands of workers, scientists and engineers, and the formation of complex organisations to construct and operate the large-scale production plants and their associated communities. Marshall formed the Protective Security Section in June 1942 to oversee aspects such as personnel, plant and military information security.

Captain John Lansdale, Jr. was drafted in in June 1941 to become part of the Investigations Branch in the Office of the Assistant Chief of Staff (G-2) and in February 1942 had been charged with responsibility for counterintelligence at Rad Lab in Berkeley and later Marshall brought him into his Protective Security Section. By early 1943, the pace of development of the Manhattan Project would ensure that G-2 was expanded with full access to the intelligence and security facilities of the Army service commands giving it the foundation for a countrywide permanent organisation with eleven branch intelligence offices at key points across the United States, from New York to Pasadena. The most serious espionage activity came from America's wartime ally the Soviet Union. Soviet agents, masquerading as diplomatic and consular officials, turned to members of the Communist Party of the United States (CPUSA) and to party sympathisers for assistance.

Groves took on the daunting task of maintaining security by insisting right from the start on a policy of strict compartmentalisation whereby contact between the various Manhattan Project sites be kept to an absolute minimum. He wanted each scientist to know only what was necessary to carry out the roles they had been allotted. The scientists, especially Oppenheimer, found this unnecessarily constricting and wanted a free flow of information between sections. It was common practice in their world for free exchange of information to stimulate new ideas and make connections between disparate fields of research. Both men were right, of course, but exactly how both innovation and security might coexist was a problem that evaded solution. For his part, Groves kept a tight grip on his own department by operating with a tiny staff of carefully chosen people and insisting upon rigid discipline. All office documents were locked away each night in a triple-combination safe. All public mention of uranium was banned. There was widespread monitoring of telephone calls. Mail was intercepted and read and some staff were put under routine surveillance.

Groves had been told from the start that the Soviet Union had an ongoing operation to discover the secrets of US fission research and he was instructed to maintain a strenuous counterintelligence programme.

The well-known predilection for left-wing politics amongst the scientific contingent was a constant worry for him. Some staff firmly believed that the US should be open about its nuclear research and share it, especially with the Soviets, while others had grave doubts about the morality of developing nuclear weapons at all. It was becoming apparent that the German nuclear programme was faltering, which weakened the argument for the US developing the bomb. At Berkeley, Army undercover agents were infiltrated onto the staff of the Rad Lab in April 1934. Elsewhere in San Francisco, the Universal Subscription Company was set up as a cover for agents and a house on Forest Avenue became a listening post for telephone monitoring.

As far as acquiring intelligence about the German atomic programme, there was precious little trust or coordination between existing agencies, so in September 1943 Marshall asked Groves to assume responsibility for all foreign intelligence in the atomic field, which he did despite having a lower rank than any of the commanders of the Office of Naval Intelligence (ONI), G-2 or the OSS, all of which were willing to give him their full cooperation. With his customary speed and efficiency, Groves called for a new streamlined command and control apparatus to monitor the movements of German scientists and industrial activity that could be associated with new large-scale construction sites inside German-occupied territories. Where links to atomic bomb activity were discovered then both direct and covert military measures would be taken to slow down or eliminate such programmes.

The crucial areas to target were the uranium mines at Joachimsthal and a prominent uranium processing centre outside Berlin, using various methods of observation and analysis including aerial surveillance photographs and making full use of various underground or partisan movements. Lists were compiled and systematic checks were made on all the precious metal refineries, physics laboratories, the handlers of uranium and thorium, manufacturers of centrifugal and reciprocating pumps, power plants and other such installations. Where installations were suspected of involvement in atomic bomb activity they would become a target of the scientific intelligence mission.

Using British agencies became a fundamental part of intelligence-gathering since they had been much more active in monitoring German nuclear activity. The British were ever willing to co-operate but assumed, given their greater proficiency and experience and the fact that the US was an intelligence novice, that they would take a leading role or at least an equal one. Groves was never likely to agree to this, however. He was sceptical about the provenance of British intelligence that seemed so comprehensive and detailed about

German nuclear activity and unmanned rocket research. The British, of course, had acquired much of this intelligence from their Ultra signals intelligence coming out of Bletchley Park where German codes were being routinely broken. Groves did not know about the Enigma transcripts and was inclined to see British reports of German atomic research as little more than gossip emanating from the Tube Alloys émigré contingent who might or might not still have some contacts in Germany that might or might not be reliable.

The First Washington Conference (code-named Arcadia) from 22 December 1941 to 14 January 1942 had set the strategic direction for the Anglo-American war effort but had almost nothing to say about co-operation in nuclear research. The follow-up and hastily called Second Washington Conference (code-named Argonaut) in June did, however, touch on the subject. Specifically on the afternoon of 20 June 1942 at the Hyde Park home of Roosevelt, he and Churchill with Roosevelt's close confidant Harry Hopkins, held a private meeting of which no official record has been found. However, Churchill alluded to it in a telegram he had later sent to Hopkins when he recalled that 'the whole basis of the conversation was that there was to be complete cooperation [on the development of atomic weapons]'. In another telegram to the same man on the same subject he wrote 'my whole understanding was that everything was on the basis of fully sharing the results of equal partners'.[7] In his memoirs, Churchill writes of his 'painful' awareness of the dangers of doing nothing, knowing what efforts the Germans were making to secure supplies of heavy water. He was convinced that the Allies could not run the risk of 'being outstripped in this awful sphere' that might mean the Germans 'get an atomic bomb before we did!'[8] For his part, Roosevelt was spared the burden of labouring under any official agreement of equality in the atomic bomb project by the absence of any written record of the 20 June meeting.

Now that Groves saw what progress the US scientists had made in the first months of 1942 by translating theory into practice it worried him that the Germans could be taking similar strides in their atomic project. If it could be done by the Americans, it could be done by the Germans. On 1 June 1942, Szilard and Compton had reached similar conclusions and they warned Bush that it was becoming essential to gather information since it had 'recently become [apparent] that the threat of German fission bombs is even more imminent than we supposed [a month earlier] . . . Secret Service activity in Germany is urgently needed, to locate and disrupt their activities'.[9] The problem for Groves and Bush was that few intelligence officers in the 1940s possessed a scientific or technical education. Instead, most intelligence professionals were trained to evaluate political, military or economic data so even if they were able to infiltrate the scientific establishment of Germany they would not have the necessary background to understand the kinds of information they should be looking for. Scientific language presented serious-enough problems on its own but translating German technical data into English was a challenge for even the best-trained linguist.

In the US, technical and administrative difficulties mounted. All four production processes for fissionable material for a bomb remained under consideration but it was time to reduce the options. When Groves took command in mid-September, he made it clear that if a bomb was to be built in time choices had to be made and made quickly. He called a meeting of the Military Policy Committee for 12 November and of the S-1 Committee two days later. Lawrence felt sure that his electromagnetic method would be chosen. By now it seemed that it was plutonium rather than uranium that held the key to a bomb. The centrifuge pathway to purification of U-235 was shut down. Gaseous diffusion, the pile and the electromagnetic method were to proceed directly to full-scale, eliminating the pilot plant stage. Later in November 1942, however, a review recommended, over Lawrence's vehement objections, that gaseous diffusion be given highest priority. At one point the gaseous diffusion plants would consume between 5–10 per cent of all electrical power generated in the US

The various aspects of research were allocated. Fermi took control of experiments and standardisation of measurement techniques. Plutonium purification work was begun at the Met Lab and an engineering division was organised under the collaboration of physicists on bomb design and fabrication. Four divisions were set up: Hans Bethe handled theory, Robert F. Bacher experimental physics, Joseph W. Kennedy took metallurgy and a navy man, Captain William S. 'Deke' Parsons, had ordnance. Los Alamos was soon threatening to burst at the seams.

While construction and operation of the production facilities planned for the Clinch River site in Tennessee were turned over to DuPont and an experimental pile would be built in the Argonne Forest Preserve just outside Chicago, Oppenheimer, with his team including Felix Bloch, Hans Bethe, Edward Teller and Robert Serber, was conducting theoretical studies which showed that approximately twice as much fissionable material would be required for a uranium bomb than had been previously estimated, which was disturbing in view of the general consensus that at least two bombs would be needed.

Shortages of uranium and graphite had delayed construction of Fermi's experimental Chicago Pile (CP-1). He needed 50 tons of yellowcake but if he got it, there was every expectation that he could start a chain reaction and Fermi and his team anticipated a successful experiment by the end of the year. Seaborg was optimistic that afterwards he could separate out plutonium from the irradiated uranium. By August 1942, Fermi's team had made a major breakthrough by producing a microscopic sample of pure plutonium, but it remained to be seen whether it could be scaled up to industrial production levels useful to the bomb effort. Groves was in Chicago in October 1942 ready to force a decision on design. Szilard argued that he was ready to go while Compton prevaricated until Fermi had completed his experiments with CP-1 and finalised his calculations of neutron production. Some scientists at the Met Lab urged that a full production pile be built immediately, while others advocated a multi-step process, perhaps beginning with an externally-cooled reactor as proposed by Fermi. Groves was not a man given to delay. He gave Compton one week to come up with a plan.

Not surprisingly, Compton compromised and chose a multi-faceted programme. Fermi would forge ahead with CP-1 and report back by the end of the year while an intermediate pile with external cooling would be built at Argonne and operated until June 1943, when it would be dismantled to see how much plutonium had been produced. The 100,000-kilowatt helium-cooled pile would be built at Oak Ridge and brought into operation by March 1944 in the hope that it would function as both a test facility and the first unit of the full-scale production plant. Studies on Szilard's idea would continue also in what was, out of necessity and given the government's deep pockets, a way to explore every possibility.

Fermi's plan was to use CP-1 to work out the value of the neutron reproduction factor (k) for a theoretical reactor of infinite size. His prototype was to be built on a squash racquet court under the abandoned west stands of Stagg Field, the University of Chicago's football stadium. Compton was so nervous about the possibility of an accidental explosion that he planned to build the industrial pile at a site in the Argonne Forest in the Cook County Forest Preserve in Palos Park, a much more remote location just in case things went slightly awry, but construction was delayed by labour disputes. Fermi was able to convince Compton that there would not be enough fissile material at Stagg Field to start a catastrophic runaway chain reaction and Compton reluctantly authorised the Stagg Field pile. Approval for the experimental pile was neither sought nor given from either Groves or the University of Chicago administration.

The uranium ore for CP-1 had been processed at MIT in Cambridge, Massachusetts into uranium oxide and delivered to Fermi's laboratory. In the room was a huge oblong pile of black graphite bricks and wooden timbers. The wooden timbers supported a lattice structure that contained over six tons of pure uranium metal, along with 34 more tons of uranium oxide. The almost 400 tons of graphite bricks in the assembly were there to serve as moderators to increase the rate of fission of the uranium. What prevented a fission chain reaction from getting out of control were a series of cadmium rods inserted into the pile's side that would, if required, close down the process by absorbing the free neutrons emitted by the fissioning uranium. There was no radiation shielding and no cooling system of any kind because the reactor was to be run at very low power. Fermi had convinced Compton that he could control the process and avoid an explosion but the risk of conducting 'a possibly catastrophic experiment in one of the most densely populated areas of the nation' remained. It was a supreme act of faith on Compton's part to allow Fermi to go ahead with the experiment.

This is a good place to explain what is meant by the term 'critical' in the context of a nuclear reactor. A reaction is said to sub-critical when its nuclear fuel does not sustain a fission chain reaction, i.e. the reaction does not release sufficient neutrons to sustain an ongoing series of reactions. Despite its ominous overtones, critical is the term used to describe a reactor state in which the number of neutrons being produced equals the number being absorbed, which in turn produces the same number of neutrons. In other words, when the reactor is critical, it is operating at steady-state. If the rate of neutron production is allowed to increase beyond the rate at which they are absorbed then the reactor will go super-critical and result in a catastrophic explosion. When a reactor starts up, the number of neutrons is increased slowly in a controlled manner. Neutron-absorbing control rods in the reactor core are used to control the process. The deeper the rods are lowered into the reactor core, the more neutrons the rods absorb and the less fission occurs. Technicians pull up or lower down the control rods into the reactor core depending on whether more or less fission, neutron production and power are desired. Should a malfunction occur, technicians can remotely plunge control rods into the reactor core to quickly soak up neutrons and shut down the nuclear reaction.

Tests carried out on the early afternoon of 1 December showed that the pile was almost ready. Fermi was confident that if he withdrew the cadmium control rods the fission chain reaction in the pile would be self-sustaining. The next morning the crowd of onlookers on the balcony grew and eventually included Walter H. Zinn, Herbert L. Anderson, Szilard, Wigner and several dozen other physicists who had played a role in the pile's construction. They were grouped around an instrument console at one end of the balcony from where they could operate a set of control rods. At 09:45, Fermi ordered that three of the safety rods be withdrawn. George Weil was on the squash court floor ready to physically withdraw the final control rod and, if necessary, re-insert it. There was also an automatic control rod which would also insert itself if the reaction reached a certain pre-set level and threatened to go super-critical. As a third fail-safe measure, Norman Hilberry stood on the balcony with an axe ready to cut a rope that would release another emergency control rod into the pile and as a final contingency, the pile would be flooded by a cadmium-salt solution.

Just after 10:00 Fermi began removing control rods and immediately the counters started clicking, measuring neutron production, and Fermi watched as the rate of production levelled off. At 10:37 he ordered Weil to pull all but 13ft of the final rod out of the pile. Again the clicking of the counters began to race and again the clicking levelled off. Weil and Fermi repeated this process, 6in at a time. Over the next few hours, the pile inched its way toward criticality. The counter activity rose dramatically. Fermi remained calm and confident that that the pile was subcritical and could point to the spot on his graph where neutron production would once again level off and that is exactly what happened.

At 11:25, the intensity of the neutron production increased to the point at which an adjustment of the instrumentation scale was required. Fermi ordered the automatic and emergency control rods reinserted for a final safety check before attempting to go critical. Ten minutes later the rods were removed in order for the experiment to resume and the reactor started up again, the counters ticking wildly for a moment before rather suddenly, at 11:35, a loud noise startled those watching. The automatic control rod had come crashing down into the pile, bringing reactivity to a complete halt while the reactor was still below criticality. The safety mechanism had been set too low during the safety check and was activated prematurely. While everyone present took a few deep breaths, Fermi realised what had happened, smiled with relief and, with admirable *sang froid,* announced to the group that it was a good time to break for lunch. All the control rods were reinserted in the pile, locked in and the group braved the freezing cold to walk to the main university dining room at Hutchinson Commons. In the splendour of a glorious Gothic replica of the Great Hall at

Oxford's Christ Church College, they had a quiet lunch and spoke of anything except what they had just witnessed.

They returned to the squash court at about 14:00 and Fermi asked the team to return the safety rods up to their positions prior to the lunch break. Over the next hour or so, Weil gradually withdrew all but one of the control rods and then slowly started to remove that one too, according to Fermi's instructions. 'This is going to do it', he assured Compton, who had joined the group after lunch, with a wide-eyed Crawford Greenewalt in tow. Even as president of the DuPont company, Greenewalt had never seen anything like this. Fermi ran through some final calculations on his slide rule as the instruments clicked away and then levelled off with a steady hum, too fast for the ear to count. 'Now it will become self-sustaining', Fermi explained. 'The trace will climb and continue to climb', he said, referring to the line being drawn across the graph drum attached to the counters. 'It will not level off.' The counters picked up speed and this time the clicking became a high-pitched whine. The intensity of neutron emission was increasing exponentially by doubling every two minutes. Tension in the room mounted as the line traced on the graph paper moved upward in an ever-steepening curve. Fermi, his face motionless but calm, took some measurements. At about 15:25 Fermi ordered a full foot of additional withdrawal of the last control rod. As Weil followed Fermi's instructions Fermi completed one final calculation and his face broke into a broad smile as he announced that the reaction had become self-sustaining or in scientific terms had 'gone critical'. It was the first time in history that man had achieved a controlled release of atomic energy, albeit at a rate that was barely enough to power a small light bulb. After 28 minutes he called a halt and Zinn replaced the safety rods into the pile. At 15:53, the world's first controlled fission chain reaction came to a complete halt. Had Fermi allowed it to continue, the reactor, without any sort of cooling mechanism, would have become intensely hot and exploded, killing everyone in the room.

Eugene Wigner produced a bottle of Chianti and paper cups were passed around for everyone to drink. The scientists drank a silent toast and all but Wigner signed the bottle's label. No code had been prearranged to inform Conant and Groves so an improvised message was sent; 'the Italian navigator has landed in the New World'. 'How were the natives?' asked Conant. 'Very friendly,' was the reply. Szilard did not share in the elation and, in reference to the military implications of what had just happened, told Fermi that the day would go down as a 'black day in the history of mankind'. Hans Bethe went away with a firm conviction that 'a weapon would [now definitely] be made'.[10]

Getting a patent on their historic invention turned out to be nearly as difficult as the breakthrough itself. As far back as 1934, Szilard had applied for a patent for his concept of a nuclear chain reaction almost as soon as the neutron had been discovered. His patent, however, specified lithium as the fissile material which proved to be quite impractical. Work on Fermi's initial patent application had begun six months before the reactor was completed. As the US reactor programme grew over the next two years, the number of patents increased. The highly classified patent for the CP-1 reactor was not filed with the US Patent Office until December 1944. It listed Fermi and Szilard as co-inventors and described the method by which a self-sustaining nuclear chain reaction had been achieved. The patent (number 2,708,656) was finally issued on 18 May 1955 and included the words, 'A neutronic reactor which comprises a moderator of graphite and natural uranium rods disposed in a geometric pattern therein . . . the purity of the graphite and the uranium and the total mass thereof being sufficient to sustain a chain reaction'.

During a meeting in May 1942, Bush and the commissioner of patents Conway Coe had told William Asahel Shurcliff, a member of OSRD, about a top-secret programme to develop a new weapon, the atomic bomb, based on discoveries openly published a few years earlier. Bush was afraid that private inventors would work out what was going on and file patent applications, thereby staking a legal claim to such inventions and endanger the security of the US atomic programme. French physicists of Joliot's Collège de France team had done something similar by attempting to file nuclear reactor-related patent applications in multiple countries.

For Bush, it seemed beyond doubt that any patent applications which seemed to have significance in regard to the US atomic bomb programme should be withheld from issue. He could not challenge patents claimed by other scientists which would run the risk of exposing the government's own nuclear designs and bomb programme but neither could he allow the patent applications to succeed and make the US nuclear programme deferent to French. The solution was to declare the patent applications secret according to Public Law No. 700, meaning that patents would be allowed to sit, unprocessed, in a Patent Office vault and only after the secrecy order was lifted would they be presented to a patent examiner.[11]

Collaboration with the chemical giant DuPont had begun in the US even before Seaborg succeeded in isolating a sample of plutonium. It had been agreed with DuPont to pursue all four methods of plutonium separation with priority given to the lanthanum fluoride process Seaborg had already developed. When DuPont got into difficulties with this method in late 1942 Seaborg moved onto the possibility of using bismuth phosphate and Dupont went ahead with an experimental production facility near Stagg Field. Under the supervision of Greenewalt, DuPont began constructing the chemical separation pilot plant at Oak Ridge, while Seaborg continued refining the bismuth phosphate method.

The sheer scale of the Manhattan Project meant that total security was never going to be possible in a country that valued personal freedom as much as the US did. It was odd, however, that where espionage by foreign powers was uncovered the punishments meted out to Manhattan Project employees were not more severe. Culprits such as David Greenglass and Clarence Hiskey would be quietly removed from post and ushered out to Alaska. None of the cases of espionage of the Manhattan Project resulted in significant charges being brought and no severe punishments were handed out, not least because much of the evidence confirming their guilt was collected by illegal means.

Clarence Hiskey, whom the Soviets called Ramsey, was a chemistry professor at Columbia and an enthusiastic member of the CPUSA. In line with many immigrant families, his own had anglicised its name to assimilate into US society, in their case from Szczechowski. He had gone out to fight with other US volunteers in the Abraham Lincoln Brigade on the side of the Republicans in the Spanish Civil War. In March 1942 he believed that the Germans were far ahead in their atomic bomb research almost to the point where they were ready to deploy a weapon which motivated him to pass on information about atomic research taking place at Columbia to Zelmond Franklin who had connections with the New York *rezidentura.* Moscow doubted that Hiskey's assessment of German developments was accurate but returned to him with a comprehensive questionnaire about his work. Hiskey would go on to work with Urey on gaseous diffusion at the K-25 plant at Oak Ridge but he came to the attention of US security through their routine monitoring. When his connections to known espionage agents was confirmed, Hiskey was called up for military service and posted to Mineral Springs in Alaska where he was tasked with sorting winter underwear for the military. He was later sent to Hawaii where he was employed manufacturing soap. If Groves could not indict Hiskey he made a very good job of humiliating him.

As a result of the new enthusiasm for British-US cooperation, Akers took a party including Simon, von Halban and Peierls to the US in early 1943. The delegates were given full access to all US research projects and the whole US operation seemed 'fired with a new vigour and determination' but beneath the bonhomie this new approach under the auspices of the US Army was already beginning to look less like a partnership as millions of US dollars were poised for investment. The Americans were now strongly set against creating any sort of integrated project that would allow the British access to all the data. Britain was no longer thought by the US as being able to make any significant contribution over and above what they themselves could do. Furthermore, the security regulations put in place by the US Army led to such severe restrictions on the information exchange that the only real traffic between British and US scientists was related to gaseous diffusion and to the use of heavy water. This change in the US attitude toward cooperating with them came as a great shock to the British. Churchill took it up with Roosevelt and although he seemed to get little immediate response, later in July 1943 Bush and Stimson went to see Churchill in London. A number of misunderstandings on both sides were addressed and seemed be resolved. It was agreed that the British should draft an agreement defining the terms for future collaboration on the bomb project which would go forward for discussion in a top-level meeting scheduled for August in Montreal.

The new year found British and US collaboration stalled which puzzled the British. Amongst other things Groves was clearly uneasy about the number of foreign scientists working in British nuclear research laboratories. His approach to security in the Manhattan Project had been to compartmentalise the different departments and restrict contact between them beyond the essential. This was quite different to the British way of doing things but scientists working on the British side were still important and had to be integrated into the US programme as seamlessly as possible. Another concern was the extent to which the British government would gain access to all US technical information about the various plants that were planned or under construction and which had been financed exclusively by the US taxpayer. Allies in war was one thing but business was still business west of the Atlantic Ocean.

Conant addressed the issue with a memorandum in January 1943 that tried to restrict British access to research and development to only those areas that had relevance in the context of the current war. Any research on such as issues as, for instance, electromagnetic separation or the manufacture of heavy water that would only really become important after the war was to be withheld from them. While the British were incensed, Conant got approval for his recommendations direct from the Oval Office. If the British were not going to get information that

Lise Meitner and Fritz Hahn, at Dahlem chemical laboratory c. 1913.

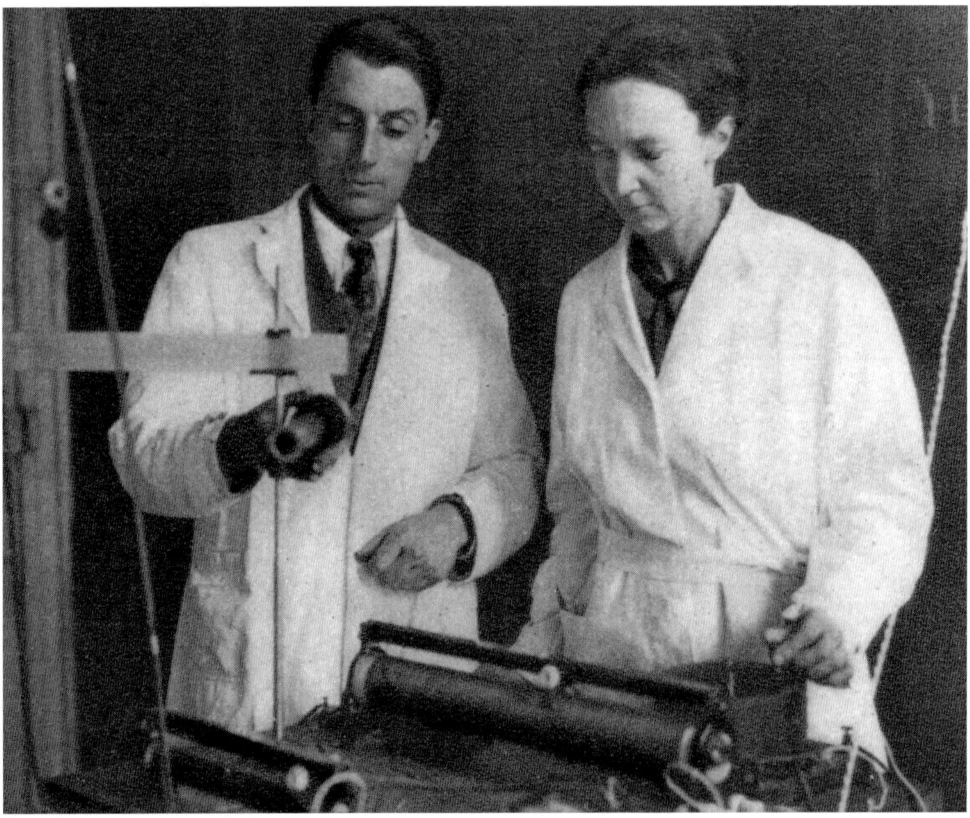

Frédéric and Irène Joliot-Curie in their Paris laboratory c. 1935.

Samuel Goudsmit (far left) at the University of Michigan in Ann Arbor in 1939, with Werner Heisenberg and Enrico Fermi (third and fourth from left).

Shinkolobwe uranium mine in the Belgian Congo.

Tiuia Muiun uranium mine in Kyrgyzstan

The French banker and intelligence agent Jacques Allier.

The Norsk Hydro heavy water plant.

The main US atomic bomb laboratory at Los Alamos which became a boomtown almost overnight.

The gaseous diffusion plant at Oak Ridge.

A British Airspeed Horsa glider of the type used in Operation Freshman.

The 'Fat Man' plutonium bomb that was dropped on Nagasaki on 9 August 1945.

The nuclear scientist Igor Kurchatov, the most important individual in the success of the Soviet atomic project.

Soviet nuclear scientists Vladimir Vernadsky and Alexander Fersman.

The first Soviet atomic bomb, 'Joe-1'.

The Japanese nuclear scientist Yoshio Nishina, a chubby man with a handsome smile and a short temper.

would help them develop their own independent atom bomb after the war, neither were they going to have access to essential raw materials.

The search for a location for the full-scale plutonium production facilities began soon after DuPont had joined the production team. The company was reluctant to get involved at first having been accused of profiteering during the First World War but its suitability for the role meant that they came under a great deal of pressure and it took a personal call from Roosevelt to the president of the company before he eventually agreed on the understanding that it would make no profit from its participation and hand over all patents acquired during the contract to the government at the end. The plutonium programme required an area of approximately 225 square miles for three or four pile reactors and one or two chemical separation complexes that would need to be at least a mile apart for security and safety purposes. Of the possible sites inspected in Washington, Oregon and California, none had a better combination of isolation and abundant water for hydroelectric power than those found along the Columbia and Colorado Rivers. It was agreed that the area around Hanford, Washington, was the most suitable and approval for its development was granted in January 1943. The Grand Coulee and Bonneville Dams provided hydroelectric power, while the flat and rocky ground was ideal to support the massive plutonium production buildings. The 400,000-acre site was far enough inland to meet security requirements, so Groves authorised the establishment of the Hanford Engineer Works, code-named Site W.

Most residents living in Hanford, White Bluffs and Richland were given 90 days' notice to abandon their homes. Homeowners were compensated based on the estimated value of their homes, excluding the value of improvements, crops and equipment. Many of the landowners rejected initial offers on their land and took the Army to court, forcing it to settle for a higher price to avoid protracted litigation. While farmers and other local residents received at least some compensation, the Native American tribes who frequented the area for fishing, hunting and ceremonial purposes received none. During the wintertime, the Confederated Tribes of the Umatilla Indian Reservation, the Confederated Tribes and Bands of the Yakama Nation and the Nez Perce Tribe lived in the Columbia Basin, while the Wanapum Tribe who lived there all year round were forced to resettle in Priest Rapids. The Manhattan Project justified its approach by saying that the area was an isolated wasteland and the people were expendable. Although they received no compensation for the land, the tribes were given visitation rights that allowed them to camp, hunt or fish in the Hanford area but this too was later revoked.

Little was known in the US about German nuclear research at the start of 1943 except that it was probably being undertaken with urgency and expertise. In Britain, Eric Welsh at MI6 had some inclination through his contact with the British spy Paul Rosbaud who had remained in Germany when his family fled, to continue his work with the scientific periodical *Die Naturwissenschaften* which kept him in touch with the scientists in the Uranverein. Through Rosbaud, Welsh drew up a list of scientists who were either working on a bomb or could make a contribution if called upon. At the top of his list was Niels Bohr in Copenhagen. Bohr had been to a very large extent free from Nazi interference in his work in a country that had not felt the stamp of Nazi jackboots quite as hard as others but he was still working in a Nazi-controlled environment and it was not known quite to what extent he may have been coerced.

Welsh knew, however, that Heisenberg and von Weizsäcker had kept in touch with Bohr from Berlin and had tried to inveigle him into getting involved with German nuclear research but as far as he knew Bohr had resisted and followed his own research paths. Welsh decided to make an offer to Bohr and sent a message through the Danish Resistance at the end of March 1943 suggesting that he escape to Britain. Bohr was cautious and doubted the authenticity of the message, fearing that it was some form of entrapment by the Nazis. Welsh followed up by persuading Chadwick to write a personal letter to Bohr that was then microfilmed and put in the hollow handle of a door key before being smuggled through Sweden to Copenhagen. 'I have in mind a particular problem in which your assistance would be of the greatest help,' Chadwick wrote.[12] Bohr obviously had a good idea to what Chadwick was alluding but had already decided that an atomic bomb was 'completely impractical' and declined the invitation but a few months later he changed his mind after the Nazis had accelerated the persecution of Danish Jews and his microfilmed acceptance was secreted in a hollow tooth of a courier.

The Nazis declared martial law in Denmark on 29 August 1943 and on 28 September, Bohr was warned that an order had been made out for his arrest. Two days later he was smuggled across the Øresund in a trawler with a dozen others to Limhamn on the Swedish coast. From there he went to Stockholm where he was joined by his wife and sons and two weeks later flown to Scotland in the bomb bay of an RAF Mosquito aircraft. He was then whisked down to London to be feted at a dinner at the Savoy Grill with MI6 chief Sir Stewart Menzies and

Wallace Akers of Tube Alloys amongst other luminaries who brought him up to date on the British atomic bomb programme. Almost immediately afterwards, Bohr and his son Aage, who had joined him, went to the US under the pseudonyms of Nicholas and James Baker while Bohr's wife remained in Britain.

Groves was now drawing on all his experience of large-scale construction projects and did not flinch from making the big decisions. On the assumption that, although Lawrence could not produce enough U-235 to be of military significance, he might be able to provide a small quantity of enriched samples quickly for laboratory work, on 9 December 1942 the S-1 Executive Committee decided to support him by constructing a small electromagnetic plant. Then on 28 December, Roosevelt approved the investment in excess of $2 billion to build full-scale gaseous diffusion and plutonium plants as well as heavy water production facilities. Bush reaffirmed his belief that mid-1945 was the earliest a bomb could be built. Full-scale electromagnetic and gaseous diffusion production plants were planned at Oak Ridge at sites designated as Y-12 and K-25, respectively, to convert new and untried laboratory technologies directly to large-scale production processes.

The vast size of these plants was illustrated by Y-12 with its arrangement of huge electromagnetic coils connected by a bus bar in an oval racetrack configuration, as seen from above. Forty-eight gaps in the racetrack between the coils would each contain two vacuum tanks. With two racetracks per building, ten buildings would be necessary to hold the estimated 2,000 sources and collectors needed to separate 100 grams of U-235 daily. It was agreed that Stone & Webster would take over design and construction of a 500-tank facility at the Tennessee site, while Lawrence's laboratory would play a supporting role by supplying experimental data. Westinghouse Electric and Manufacturing Company, the Allis-Chalmers Manufacturing Company and the Chapman Valve Manufacturing Company were contracted to provide electrical equipment. Groves applied pressure by demanding that the first plant be in operation by 1 July 1943. This was a tall order given that each of the five planned racetracks was 122ft long, 77ft wide and 15ft high, that the completed plant was to consist of three 450ft long buildings, each housing two racetracks placed end-to-end on the second floor and that separate chemistry buildings also would be needed for preparing charge materials and separating uranium recovered from the tanks. The scientists were stunned by the scale of the project but Groves maintained that his schedule was non-negotiable.

The first sod was turned on 18 February. Blueprints could not be produced fast enough to keep up with construction as Stone & Webster forged ahead. The magnets needed so much copper for windings that the Army resorted to using almost 15,000 tons of silver bullion from the United States Treasury to make the electrical coils as a substitute for copper but the most serious shortages were of electronic tubes, generators and regulators. On top of this, Oppenheimer warned that three times more fissionable material would be required for a bomb than earlier estimates had indicated. Lawrence responded by persuading Groves to build a Y-12 Extension which, in effect, doubled the size of the electromagnetic complex. All this was being done with constant modifications to a production plant that did not have the benefit of being preceded by a pilot plant to iron out difficulties beforehand. There was no shortage of problems confronting Groves but he was a man used to meeting challenges head on. Vacuum tanks in the first racetrack leaked and moved out of line due to magnetic pressure, welds failed, electrical circuits malfunctioned and operators made frequent mistakes but most seriously of all, the magnet coils shorted out because of rust and sediment in the cooling oil.

The X-10 Graphite Reactor was the first reactor built after the successful experimental CP-1 at the University of Chicago. This was a pilot plant based on Fermi's pile and was successful in producing the world's first plutonium outside the laboratory. This technology was the foundation upon which the giant plutonium-producing facilities at Hanford were based. CP-1 had been an undoubted success as a scientific experiment, but that pile was built on such a small scale that recovering any significant amounts of plutonium from it was impractical. Calculations that Fermi made as a result of CP-1 showed a higher value for the neutron reproduction (k) than had been anticipated which suggested that water cooling of the pile was now a practical possibility but Greenewalt was not convinced and continued to support helium cooling. Fermi's value for 'k' also seemed to make possible an experimental pile using air cooling which would be easier to design and quicker to build. DuPont started the construction of X-10 in February 1943. First, a series of huge underground concrete cells were built then covered by an enormous concrete shell 7ft thick with hundreds of holes for uranium slug placement. It was ready to run its first tests by 4 November, producing its first plutonium by the end of the month. Gradually the power would be raised and the plutonium yield increased.

The Met Lab now occupied a less central place in the bomb project as Oak Ridge and Hanford rose to prominence. Fermi continued to work with CP-1 but experiments at the Argonne site using CP-2, built with material from CP-1, began to take precedence as Met Lab research scientists found themselves serving primarily as consultants for DuPont's work at Oak Ridge and Hanford, where plutonium production facilities took shape with some urgency. The three water-cooled piles (designated B, D and F) were to be built about six miles apart on the south bank of the Columbia River. The four chemical separation plants would be built in pairs at two sites nearly 10 miles south of the piles. A facility to produce uranium would be approximately 20 miles south-east of the separation plants near Richland.

Work began by Stone & Webster in June 1943 to design the Argonne Forest pile but it soon became clear that it would meet with problems not encountered in Fermi's experimental pile. The high power levels of Compton's pile would mean that radiation and containment shielding and an efficient cooling system would be essential. The proposed design called for a 460-ton cube of graphite to be pierced by 376 vertical columns containing 22 cartridges of uranium and graphite. It would be cooled by circulating helium from top to bottom through the pile and a shield of graphite surrounding the reactor would provide radiation containment. An alternative to this design was one proposed by Szilard based on one that he had worked up with Einstein using bismuth in liquid metal form but the cost weighed heavily against it.

The Maud report of 1941 had been in favour of gaseous diffusion as the most promising method of enriching uranium and the 1942 Lewis committee placed it first among isotope separation methods. It was an act of faith therefore when, despite the soundness of the theory, the method had not produced a single iota of enriched uranium when the K-25 Gaseous Diffusion Plant was authorised. As a consequence, it was decided in late summer 1943 that K-25 would not attempt to create weapons-grade uranium but instead would produce material that was about 50 per cent U-235, which would then be fed into Y-12's Beta tracks for final enrichment. This downgrade of K-25 was part of the larger decision to double Y-12 capacity in accordance with Groves' strategy of using both methods to produce enough fissionable material for bombs in the shortest possible time. The level of resources that had been poured into K-25 meant that Groves really had no option but to take this view and hope that Y-12 would survive its growing pains and that maybe the issues with corrosive gases in K-25 would be overcome before too long. In the summer of 1944, there were serious doubts if a bomb could be built before the end of 1945.

As for K-25 (Kellog-25), finding a suitable location for the full-scale gaseous diffusion uranium production plant at the Clinton Engineer

Works in Oak Ridge soon came into question. Du Pont highlighted the hazards of producing plutonium on a large scale and Groves worried about placing the facility near to the electromagnetic diffusion plants. Furthermore, the site was far too close to Knoxville should a catastrophe occur. Lack of power generation was not the least of their worries, so the search began for an alternate location for the full-scale plutonium plant. The planned three or four plutonium production reactors and one or two chemical separation complexes would need to be at least a mile apart for security purposes and nothing could be allowed within four miles of the separation complexes for fear of radioactive accidents, and roads, railways and laboratories would have to be even further away.

Greenewalt was worried that the water-cooled design for Hanford would lead to water boiling in some of the tubes, reducing the safety margin, but he was even more worried about the proposed helium-cooled model. He feared that the compressors would not be ready in time, that the shell could not be made vacuum tight and that the pile would be extremely difficult to operate. DuPont engineers conceded that Greenewalt's fears were well-grounded but late in February, they reluctantly concluded that the Met Lab's model, while it had its problems, was superior to DuPont's own helium-cooled design and decided to adopt the water-cooled approach. With this the X-10 Graphite Reactor was born.

Meanwhile, Met Lab activities were focused on the design of the proposed full-scale water-cooled piles to be built at Hanford. Taking their cue from the DuPont engineers, who utilised a horizontal design for the air-cooled X-10, Met Lab scientists abandoned the vertical arrangement with water tanks, which had posed serious engineering difficulties. Instead, they proposed to place uranium slugs sealed in aluminium cans inside aluminium tubes. The tubes, laid horizontally through a graphite block, would cool the pile with water injected into each tube. The pile, containing 200 tons of uranium and 1,200 tons of graphite, would need 75,000 gallons of water per minute for cooling. Buckets of irradiated uranium would be allowed to undergo radioactive decay for several weeks then be moved into the chemical separation facility where the plutonium would be extracted with remote-controlled equipment. On 4 November 1943, X-10 went critical for the first time at Oak Ridge. By the end of the month, it was producing small but experimentally valuable samples of plutonium ahead of the large-scale operation of the DuPont plant at Hanford. This was important work because the plutonium from X-10 was found to behave differently to the plutonium produced in a cyclotron because of contamination by the isotope Pu-240. Calculations indicated that above a certain level of contamination with Pu-240 the plutonium in a trigger-activated bomb would be susceptible to pre-detonation. So for a plutonium bomb only a relatively unexplored implosion method could be used. This involved symmetrical shockwaves being directed inward to compress a subcritical mass of plutonium into a smaller, now-critical sphere.

Oppenheimer spent the first part of 1943 scouting for staff and even persuaded Bacher and Isidor I. Rabi of the Massachusetts Institute of Technology's Radiation Laboratory to sign up. Almost overnight Los Alamos became a boomtown, as scientists and their families moved in along with all manner of experimental equipment. There was also a sizeable contingent of British scientists who helped to make up a remarkable collection of talent and machinery. Early work on the design of the atomic bomb began as scientists arrived at Los Alamos during 1943. While the properties of uranium were reasonably well understood and those of plutonium were being unravelled, knowledge of fission explosions remained quite theoretical. Seaborg's team had proven in March 1941 that plutonium underwent fission but it was not known yet if plutonium released secondary neutrons during bombardment. Theory put plutonium in the category of fissile bomb material but only experiments could prove it. The optimum size of the critical mass remained to be established, as did the optimum shape. The main question, however, was what was the optimum critical mass to release the greatest possible amount of energy before it was destroyed in the explosion?

Once the critical mass was calculated it was possible to think about a mechanism to detonate it. Two sub-critical masses of fissionable material would have to come together to form a supercritical mass for an explosion to occur. Furthermore, they had to come together in a precise manner and at high speed, at least 3,000ft per second. The only gun the US army had capable of firing a projectile weighing around 20kg with such high muzzle velocity itself weighed about five tons. Any bomb would require two such artillery pieces pointing directly at each other and primed to fire at exactly the same time. Timing was of the essence. The uranium would fission in a millionth of a second and vaporise everything in the bomb mechanism which would expand explosively and reach 'second criticality' when the chain reaction would shut down. A conventional artillery method of firing one subcritical mass into the other seemed to be the answer for U-235, but this method would only work for 100 per cent pure plutonium. A 'gun' mechanism was designed for uranium but if pure plutonium, which it was believed required a much smaller critical mass to detonate, was not available the bomb designers would be forced to look at the relatively unknown implosion method.

Using this method, the bomb would be triggered by an implosion mechanism, in which a spherical assembly made up of as many as thirty-two conventional high explosive charges compressed a subcritical core of a plutonium compound to supercritical density. It would be absolutely essential for all of the surrounding explosions to occur with synchronicity of a millionth of a second to ensure that the fissile core was compressed symmetrically. The intensity of the explosions would need to be sufficient to compress the core within a tiny fraction of a second to avoid pre-detonation. The heavy outer layer would reflect neutrons back into the active volume and contain the explosion for a few crucial moments to increasing the efficiency of the blast. If the sub-critical masses came together too slowly they would still explode but well before any chain reaction could take place and the explosion would be very much less powerful than an atomic bomb.

Groves had decided to operate Los Alamos as a military laboratory and, while Oppenheimer accepted his rationale for this arrangement, he was afraid that his scientists would be uncomfortable working within such an environment and indeed it had taken some effort to get Bacher and Rabi to join the Los Alamos team. Groves was ready to compromise and promised that the laboratory would remain civilian through 1943 but heightened security protocols would be essential during the final stages of the project.

It was about this time that Groves became aware of Sengier's uranium ore stocks on Staten Island when the latter had applied for a licence to ship it to the Eldorado refinery in Ontario for processing into yellowcake. The OSRD S-1 Executive Committee recommended that they purchase the whole stockpile from Sengier, which it did, but it was still not enough to meet all the requirements of the gaseous diffusion plant at Oak Ridge and the Hanford reactors. By this time there was a US military presence in the Congo. Nazi influence had waned in Africa as a whole but there was still a considerable stockpile of untreated low-grade uranium ore at Shinkolobwe. The Bureau of Economic warfare agent in Léopoldville, Hickman Price, was ordered to buy up all available stocks and ship them out to the US and ensure that they did not fall into German hands. While the ore, with a uranium content of less than 20 per cent, was of far lower quality than that already exported, it was still much better than anything that was coming out of the Eldorado mine in Canada. Price also requested that the mine be opened up again but by this time all the galleries had been flooded.

The ore would normally be shipped out through the port of Lobito Bay in Angola but that was one of the few places in Africa that was still 'seething with Nazi spies' so it was decided instead to bring it out through the only Congolese outlet to the sea at Matadi. This avoided having it leave the country, but the 1,500-mile route was long and difficult. It would go by rail initially, but the sealed barrels marked 'Special Cobalt' had to be hand-loaded onto barges as Port-Franqui on the Kasai river where they were sailed downstream to where the Kasai joined the Congo river and offloaded at Léopoldville onto another train for the final leg of the journey to Matadi. That was not the end of problems since the Atlantic was now patrolled by German U-boats. To avoid the catastrophic loss of the whole shipment it was loaded onto the SS *Talisman* and SS *Titania* and several other several other ships of the Barber West African Line, two of which were sunk, one by torpedo and one by accident. One-fifth of the whole consignment went to the sea floor. Not all went by sea, however. Some of it was flown out of Léopoldville and Elizabethville airports, specially upgraded with concrete runways, by aircraft of the Pan American line. Although the British also had a presence in the Congo at this time, Price was able to move this amount of ore out of the country without them being aware of it. Clearly the US was staking a claim to the whole Shinkolobwe output and was not keen for the British to know about it ahead of the Quebec Conference scheduled for August 1943.

The remoteness of the Shinkolobwe mine from Kinshasa and the existence of pre-existing export routes to Lobito Bay in Portuguese Angola made the smuggling of uranium ore to Europe a constant worry for Groves. British Intelligence had known that German Intelligence agents had been active in the Congo ever since 1940 trying to find out if the Allies were 'developing new [mining] industries or sources of raw materials'.[13] The US was naturally interested in other important minerals and diamonds sourced from the Congo and tried to disguise its interest in uranium ores to avoid giving any clues about its nuclear programme to German spies. Smuggling of diamonds in particular was rife, with tales of stones being spirited out to Germany in Red Cross parcels. British Intelligence, being generally uninformed about Tube Alloys and the Manhattan Project, were at a loss to know why the US was so interested in the Congo and why a number of German agents they had captured indicated during interrogation that they had connections with that country. It was assumed by Groves that any smuggling of raw materials from the Congo was done through Belgian companies working for the Nazis and the only way to find out more about it was to make contact with smugglers, posing as Belgian or German agents. Henry Emil Stehli, a Swiss-born member of the Stehli Silks Corporation of Zurich, had joined OSS and was sent to the Congo ostensibly to establish cover as a buyer of raw silk. He took charge of all OSS operations in Léopoldville and let it be known that he was interested in obtaining illicit industrial diamonds, assuming that the same people who dealt with that were smuggling uranium ore as well but by this time German influence in Africa was almost non-existent and the main problem was to prevent the British getting hold of any ores. The British, themselves, through the medium of *The Times* newspaper that was generally thought to be the propagandist for the British government, tried to divert attention away from their own interest in the Congo by referring to Canada as leading the Empire and the world 'in the supply of [uranium]'.

Construction began on the B Reactor at Hanford on 27 August 1943 but work on the two massive, scaled-up chemical separation plants stalled due to labour shortages and the lack of final blueprints. It was not until 10 October that DuPont engineers were able to start marking out the site for the B-Reactor pile building at Hanford. The area immediately under the pile was excavated and carefully load-tested. Assembly of the pile started in February 1944 and by spring, a windowless concrete monolith towered 120ft above the desert. The cast-iron base and the thermal and biological shields around the pile were completed by mid-May. Another month passed before the graphite blocks were installed and it took a further two to wire and pipe the pile and connect it to various monitoring and control devices. By July 1944, the B Reactor would be nearly finished and the second D Reactor about half-finished also. Work on the third F was yet to begin. Work started on the first chemical separation building, T Plant, in January 1944 and along with a similar U Plant, were completed by October also at Hanford. The B Separation Plant would eventually be completed in February 1945. It would be 800ft long, 65ft wide and 80ft high containing forty process pools. Technicians operated remotely behind concrete walls 7ft thick as precautions against radiation exposure.

Canada had built up a respectable reputation for nuclear research and had been conducting, albeit unsuccessful, experiments with uranium since 1940. When the Tizard delegation visited the US in 1940, they had made contact with George Laurence of the National Research Council of Canada that resulted in a $5,000 investment from ICI for continuation of his research. The mountains of British Columbia gave the British a further motivation for collaboration with the Canadians in the form of the Consolidated Mining and Smelting Company (Cominco) who had been involved in heavy water production there for a number of years. The NDRC had offered Cominco a contract to supply 900kg of heavy water which would have meant building an extension to their plant but Cominco were reluctant to agree since they only had a half share in the manufacturing patent and there was no guarantee of orders beyond the end of the war. When agreement was reached on the basis of a $20,000 investment by the NDRC, it was agreed informally that the British would also have access to heavy water for their research as well as uranium ore from the Eldorado mine but that was an agreement that evaded scrutiny.

The British had, by the end of 1942, come to the conclusion that it would be advisable to relocate their atomic research to Canada, both to be closer to the US programme and because Canada could supply vital raw materials. Agreement was quickly reached and the Montreal Laboratory was established to continue the work of Tube Alloys. The first eight staff arrived from Europe and began investigating reactor design. Despite Canadian assurances that the British would be able to buy uranium ore it transpired that all stockpiles had been shipped to the US and all future production contracted out to them. A similar blow was felt when it transpired that Cominco had already been contracted to sell all of its heavy water to the US.

Churchill raised the matter with Roosevelt when they met in Washington in May. The US President gave his personal assurances that Tube Alloys would be given full access to the Manhattan Project but this was quickly undermined when it became clear from discussions between officials of both sides that it was the intention of the British to become a nuclear power in their own right after the war. This did not fit in at all with how the US saw the post-war world in which the British Empire would be dismantled in favour of a dominant US as the only nation with the atomic bomb. Roosevelt was content to see the impasse drag on.

A top-secret military conference (code name Quadrant) was held in Quebec on 17–24 August 1943 in which the governments of Britain, Canada and the US discussed coordination of efforts to develop nuclear weapons. Churchill and Roosevelt signed the Quebec Agreement stating that the nuclear technology would never be used against one another and that one would not use it against third parties without the agreement of the other. The Canadian premier Mackenzie King played only a peripheral role in discussions, but his country was important because of its uranium deposits and heavy water plant.

It had been the Tizard mission in 1940 that had kick-started British-US military cooperation. The US had played hardball at the time by demanding access to all British research in exchange for a promise to do more to help the British war effort at a time when US public opinion was strongly isolationist. The results of negotiations at that time were that the US was given long leases to take over British naval facilities in the Caribbean as well as full disclosure on the cavity magnetron, a revolutionary development in radar technology. The cavity magnetron would become the bedrock of much innovation after the war, including microwave ovens, which made it hugely important financially. Unfortunately, the British, with their single-minded efforts to prevent a German invasion in 1940, had failed to patent it. The American RCA company took full advantage and reaped a huge bonanza. In return, the British had been denied access to the Norden Mk XV bombsight, which they coveted, and got only fifty rusting hulks of obsolete First World War destroyers instead. Churchill had gambled that the price had been worth it to bring the US political, industrial and military elites out of their isolationist cocoon and face up to the reality of the Nazi menace.

Almost all groundbreaking technological advances in Britain in 1942 came under Tube Alloys which was the focal point of all nuclear research. The problem for Britain, of course, was that it was mortgaged up to the hilt, under aerial attack by the Luftwaffe and desperately short of modern industrial facilities. Churchill realised straight away that if the Allies were to produce an atomic bomb before the Germans it could only be done with the vast resources of the US and the US would get there much quicker if it had full access to all British research that had been done up to that date. If he had assumed, however, that Britain would be entering into an equal partnership with the US in the development of nuclear technology, he should have remembered how, from a position of strength, the US had outmanoeuvred and exploited the Tizard mission almost with contempt in 1940. Britain had every ambition that it would become a nuclear power in the post-war world alongside the US and never lost sight of that but after 1942 it realised that by putting all its resources into Tube Alloys which would then be absorbed by the Manhattan Project it would lose all independent control of its own bomb and become totally reliant on the US. Although a working arrangement was agreed it left Tube Alloys in a decidedly junior role, unable to pursue substantive work at home and allowed access to the Manhattan Project only in areas chosen by the Americans. This economic subjugation of Britain was no accident; it was part of Roosevelt's post-war ambitions to see the already moribund British Empire dismantled and replaced as a world power by the US. An essential aspect of that hegemonic vision was that the US would be the sole nuclear power in the world.

Over the following two years around sixty scientists went from Britain to work on the Manhattan Project, making many substantial contributions to electromagnetic separation work at Berkeley, California and Oak Ridge, Tennessee and at the main bomb laboratory at Los Alamos. Peierls, Fuchs, Frisch and others all did important work relating to the design of the plutonium implosion weapon later used over Nagasaki. At the same time, May and several dozen British scientists, along with von Halban and Kowarski, joined a large British-Canadian reactor development project based in Montreal. Research and development in Britain all but ceased. The French émigrés had been greeted very coolly by Groves in Chicago who was against any foreign nationals working on US defence projects and who was glad to see the back of them when they were transferred to the Canadian mission.

As early as 30 July 1942, an agitated Sir John Anderson, the minister responsible for Britain's Tube Alloys project, had told Churchill that the pioneering British work on atomic bomb research was 'a dwindling asset' and unless Britain took steps to capitalise on it quickly it would soon have little or none left. It was clear that intellectual traffic with the US was going to be very much in one direction and so in response Anderson was instructed to approach C.J. Mackenzie, President of the National Research Council of Canada, to set up a joint Anglo-Canadian Nuclear Laboratory in Montreal. Unfortunately, while Mackenzie saw Bush and Conant as 'fair and reasonable men', he looked on senior British scientific administrators as 'arrogant, ill-informed and naïve' and who had 'a tendency to overemphasize the importance of their contribution as compared to that of the Americans'. He did, however, consider Akers to be 'very bright . . . scientifically sound and a first class engineer with [his] feet definitely on the ground'.[14]

Whilst Anderson saw no alternative but to shift the British U-235 fast-neutron team to the United States, where it would be incorporated in the US programme, the Cambridge slow-neutron team led by von Halban that was concentrating on production of plutonium could easily be moved to Canada. On 10 November 1942, a list of twenty-six Tube Alloys scientists and technicians destined for Montreal was submitted to Canadian immigration officials. Eighteen were British citizens with the remainder refugee scientists from a variety of backgrounds. Akers coordinated the activities of the Montreal Laboratory with the British fast-neutron team in Los Alamos and established a working relationship with the newly established Manhattan Project. Very soon, however, friction built up between Mackenzie and von Halban, whose had a notoriously short fuse, over procedures and priorities. There were delays and obfuscations over equipment and facilities but especially with slow procurement of vital American uranium oxide, uranium metal, heavy water and graphite In his diary, Mackenzie called von Halban 'a mere child and a temperamental one at that', an opinion shared by many within the Montreal Laboratory.[15] Akers on the other hand was frustrated with Mackenzie's 'sycophantic posturing towards American scientific mandarins'.

By the autumn of 1942, von Halban was in Montreal working under Mackenzie and bringing together British, Canadian and French scientists with the goal of furthering studies on chain reactions with heavy water as a moderator but the idea of integrating research with the US ran into difficulties because the US had chosen to work with graphite instead. At the end of October 1943 morale was low at the Montreal Laboratory with many senior British scientists openly talking of wanting to return home. Akers and Anderson conceded that they were destined to have no more than junior-partner status on terms dictated by the US. Full interchange of information between Allied scientists was confined to those 'engaged in the same section of the field' and 'between members of the Combined Policy Committee and their intermediate technical advisers'. On 7 September, Groves made it perfectly clear that the agreement would not include operational collaboration and whether they liked it or not, British and Canadian scientists would have to accept his rules. After a campaign to undermine his position by Bush and Groves, Akers was fired and replaced as scientific director of the Tube Alloys project by Chadwick. It was the end of a road that Akers had not wanted to take in the first place. His disagreements with Mackenzie, despite their relationship starting out so well, his unpopularity with many of the leading British scientists and, crucially, his acrimonious relationship with Bush and Groves had sealed his fate. One suspects that he dropped out with some sense of relief. Chadwick entered the fray with a recognition that compromise, not confrontation, was the only viable approach in dealing with the Americans since they clearly had the upper hand and could probably get along if necessary without the UK, while the UK could do nothing without the US. Akers' involvement with ICI and the sense that US research was being creamed off for the benefit of British industry could well have been the reason that the Americans were glad to see the back of him. Whatever the truth of that, Chadwick's appointment and well-known opinion that 'Britain's nuclear venture could not afford to indulge scientific pipe dreams', was rewarded with a new expression of willingness to provide British scientists with full integration into the Manhattan Project.[16]

Conant welcomed the British-Canadian initiative but wrote to Mackenzie saying that 'The United States was now committed to a new atomic exchange policy, based on the principle that we are to have complete interchange on design and construction of new weapons and equipment only if the recipient of the information is in a position to take advantage of it in this war'. His assertion that Britain clearly did not have the resources to produce its own bomb without US aid surely meant that Britain was to be denied access to US research and would be cut adrift after the war. Indeed, Conant specifically said that 'we have been directed to limit the interchange accordingly'. As far as Canada was concerned, Conant saw a role for Canada in 'carrying out the fundamental scientific work for the use of heavy water so that the DuPont Company could base their designs on this experience' but it would not be given access to 'the methods of extraction of plutonium nor the design of the plant for the use of heavy water for this purpose, nor the methods of preparing heavy water'. Conant ended his letter by expressing his 'confidence' that Mackenzie would soon recognise that these Manhattan Project guidelines were a good thing, which would 'greatly enhance the Anglo-American-Canadian objective of producing a weapon to be used against our common enemy in the shortest time under the conditions of maximum secrecy'.[17]

As a result of the Quebec Agreement, the British handed over all copies of their recent work and, in return, Groves supplied a copy of the progress report he had just submitted to Roosevelt. The scale of progress made in such a short time by the US was impressive, especially the financial investment, and could only confirm in British minds that under the circumstances the first duty of the British was to assist the US with its project and abandon all ideas of a wartime project of their own. Chadwick decided that this would best be achieved by sending British scientists to work in the US so before the end of 1943, Chadwick, Peierls and Oliphant had all taken up residence there. Chadwick spent his time in Washington on diplomatic and administrative duties while Peierls worked on gaseous diffusion and later at Los Alamos. Oliphant went to Berkeley to work with Lawrence's electromagnetic team and the British contingent here eventually totalled some thirty-five. The first of nineteen British scientists to go to Los Alamos included Frisch and became involved with implosion weapon problems and bomb assembly. Eventually six British scientists became the heads of their joint groups and a seventh became head of a section.

The Montreal Laboratory was never really taken seriously by the US team but they had to maintain good relations to ensure that they were given maximum access to Canadian uranium deposits and the heavy water plant in British Columbia. The fact that nothing being done in Montreal would advance the making of the bomb before 1945, that there was a risk it might exploit US research and development for post-war military and commercial advantage and the security aspect of its substantially foreign scientific community, especially von Halban, was suspect, all of which reinforced Groves' reluctance to share developments. As a parting shot, Akers had said in a barely disguised understatement that 'Montreal seems to have been treated in a distinctly illogical manner'.

Two US scientists, H.W. Watson and John Huffman, along with Major Horace Benbow, a member of Groves' special security team, were appointed to the staff at Montreal which essentially sealed US control and prevented any leakage of US plutonium secrets. US concerns over security standards at the Montreal Laboratory and the activities of some Montreal scientists when visiting US nuclear installations were a real issue for Benbow who discovered that many of the British scientists working on the project had 'never been [properly] cleared by British Intelligence' and all personnel records and clearances were unavailable to the Royal Canadian Mounted Police. For their part the British and Canadian scientists found the US security people 'vexatious and insulting', especially when they blocked entry to research facilities for Niels Bohr. US fears were vindicated later, however, in January 1945 when Groves would receive a report from the Argonne Laboratory about the activities of physicist Alan Nunn May and the visits made by von Halban to see Joliot in now-liberated Paris.

CP-1 had shown that nuclear chain reactions could be controlled but this was now scaled up to the X-10 Graphite Reactor (Clinton Pile), an experimental air-cooled production pile that would help in designing the full-scale helium-cooled reactors that were also being planned. The X-10 Pile was built at the Oak Ridge but the helium-cooled reactor was planned for Hanford, both built by DuPont. On 4 November 1943 the X-10 went critical for the first time and by the end of the month was producing small but significant samples of plutonium.

Implosion research found itself a little behind the gun method which meant that the man tasked with developing it, Seth H. Neddermeyer, was left rather isolated but that did not prevent him from making a significant breakthrough. Initially his work had stalled due to the difficulty in achieving a symmetrical implosion but when the Princeton mathematician John von Neumann visited Los Alamos late in 1943, Neddermeyer got him to look at the problem from a different angle and he came up with a suggestion that high-speed assembly and high velocities would prevent pre-detonation of the plutonium bomb and achieve more symmetrical explosions. Less fissionable material would be required, bombs could be ready earlier, and extreme purification of plutonium would be unnecessary. This put Neddermeyer's research back into the fast lane.

Early experiments showed that uranium emitted neutrons in less than a billionth of a second, just enough time, in the world of nuclear physics, for an efficient explosion. Emilio Segrè discovered in December 1943 that uranium of less than previously-thought purity would work, meaning that the muzzle velocity for the scaled-down artillery piece could be lower and the gun could be shorter and lighter. Furthermore, his tests on plutonium indicated that it emitted even more neutrons than uranium. All theory and experimentation now agreed that a bomb using either element would detonate if it could be designed and fabricated into the correct size and shape but many details remained to be worked out, including calculations to determine how much U-235 or plutonium would be needed for a critical mass.

Much of the work that led to the atomic bombs was extremely tedious. For instance, the chemists had to purify the U-235 and plutonium to ensure that they were pure enough to avoid premature detonation then the metallurgy division had to turn it into metal. The ordnance engineering division had to design two artillery pieces of relatively standard specifications except for their extremely light barrels – one for a uranium bomb and one for a plutonium bomb. These had to achieve high muzzle velocities, but they would not have to be durable since they would only be fired once. The plutonium weapon required a higher velocity due to its higher risk of pre-detonation and two guns were successfully field tested in March 1944.

Groves stormed in on 15 December and shut down the Y-12 electromagnetic racetrack which had been giving a lot of trouble at Oak Ridge. The coils were sent to Allis-Chalmers for cleaning. A second racetrack fared little better when it started up in mid-January 1944. Performance was sporadic and maintenance could not keep up with electrical failures and defective parts. By the end of February, however, there was about 200 grams of 12 per cent U-235 available for experiments at Los Alamos. Complete redesign of some parts of the plant meant that the racetracks came on line four months late and failed to live up to expectations when they did so. Lawrence, however remained convinced that, with K-25 experiencing even bigger problems, Y-12 still offered the only realistic avenue to a bomb in time for use in the war and Groves agreed to extend Y-12 with the construction of a third building containing another two racetracks.

Fears of a German atomic bomb persisted and so Oppenheimer convened a meeting on 31 December to which he invited Bohr, Hans Bethe, Edward Teller and others. The issue was a drawing that Bohr had brought with him to the US that he claimed was made by Heisenberg at their September 1941 meeting. Bohr displayed his poor understanding of technical matters by assuming that it was a sketch of an atomic bomb that the Germans had been working on. As soon as the others at the meeting saw it they were convinced that it was a very primitive sketch not of a bomb but of a reactor but Bohr insisted that Heisenberg had indicated to him that it was indeed a bomb. Furthermore, discussion led to the conclusion that it was not even an efficient reactor but in fact a reactor bomb that would explode long before a chain reaction could take place. As a weapon of war it was virtually useless.

While experiments on the gun and implosion methods continued at Los Alamos, Parsons directed much of his effort toward developing bomb hardware, including arming and wiring mechanisms and fusing devices. Working with the Army Air Force, Parsons's group developed two bomb models by March 1944 and began testing them with B-29s. 'Thin Man', named for President Roosevelt, utilised the plutonium gun design, while 'Fat Man', named after Winston Churchill, was a plutonium implosion prototype.

The US Navy had been kept out of what was essentially an Army operation but when both Y-12 and K-25 were unable to overcome a number of problems, the Manhattan Project was the unexpected recipient of a naval lifeline. The navy had been primarily interested in research on atomic power as a propulsion mechanism for submarines and had been ignored by Groves in his programme to build a bomb which made it all the more surprising when Groves read the contents of a letter he got from Oppenheimer in late April 1944 saying that he had made 'a terrible blunder'. The letter contained information about thermal diffusion experiments conducted by Abelson at the Philadelphia Naval Yard which were worthy of a closer look. Abelson was building a plant to produce enriched uranium and Oppenheimer thought it might be useful as a feed for Y-12 until problems with K-25 could be resolved.

Abelson had taken his research on liquid thermal diffusion to the Naval Research Laboratory because both Bush and Conant could not see it being of use within the time frame they had set for a bomb, especially since the electromagnetic and pile projects were doing reasonably well. They did not abandon Abelson entirely, however, and they encouraged the Navy to increase its support for Abelson's work even though they considered that it would never replace gaseous diffusion which was way out ahead in the popularity stakes. Groves saw straight away that Abelson might well be the answer to his problems with Y-12. It would be a relatively simple matter to build a thermal diffusion plant using steam that was already on line for K-25. With his customary zeal and no-nonsense approach, Groves contracted H.K. Ferguson Company of Cleveland to build the plant in just 90 days.

On 14 May 1944, after CP-3 had reached critical mass, Groves and Conant questioned the rationale for building a heavy-water competitor plant at Montreal, especially if it threatened US plutonium secrets. The only hope that Montreal would survive was if it could exhibit 100 per cent commitment to the US programme. Groves promised collaboration on engineering research and development as well as physics and chemistry and that Montreal scientists would be given information about the technical process for the extraction of U-235 but the price demanded was the replacement of von Halban by John Cockcroft. Conant was not mollified and was openly dismissive of the Montreal laboratory which he saw as being beyond the requirements of the US wartime effort and strictly a post-war project but Groves kept the bigger picture in mind and realised that he needed all available resources including the British-Canadian team and that meant employing the sort of diplomatic sensitivities that seemed to be beyond Conant. For a bluff army man, Groves was to display surprising subtlety in his dealings with people.

Groves' efforts were rewarded on 8 June 1944 when arrangements for the Anglo-American-Canadian heavy water project at Montreal were concluded. The deal consisted of six major points:

- to maintain existing levels of production for heavy water;
- to continue programmes at both Chicago and Montreal 'for the development of fundamental information on heavy water piles;
- to undertake the design and construction of a heterogeneous heavy water pile 'of about 50,000 KW' at Montreal as a joint project;
- to review the status of the Montreal plant 'when the performance of the first large scale graphite pile at Hanford could be assessed';
- to create the necessary administrative machinery to supervise properly the Montreal pilot pile project;
- to strengthen the manpower resources at Montreal 'by the inclusion of Canadian scientists as well as British and American scientists'.[18]

Bohr met Churchill on 16 May 1944 at Downing Street. Churchill, preoccupied with last-minute planning of the D-Day landings, had not wanted the meeting at all but felt obliged to see Bohr whom he felt he had to humour because of his unofficial role as liaison between Britain and the US. Bohr immediately ruffled Churchill's feathers by suggesting that all atomic research should be shared with the Soviet Union. Bohr was clearly not familiar with Churchill's anti-communist sentiments. 'I cannot see what you are talking about', Churchill angrily told his guest, wishing to be rid of him. Where Bohr was failing to progress his agenda of openness with the Soviets, Fuchs would be rather more successful.

A few days after this meeting, Bohr tried to mend fences with Churchill in a letter which began,

> It was . . . a revelation to me, on my arrival in England . . . to learn with what courage and foresight the [joint nuclear] effort had been undertaken and what an advanced stage the work had already reached.
>
> I was expressly asked by the leaders of the American organisation to convey their genuine appreciation of the help they are receiving, on an ever increasing scale, from their British collaborators.
>
> These circumstances obviously have an important bearing on the question of an eventual competition about the formidable weapon and on the problem of establishing an effective control and might therefore perhaps influence the judgment of the statesmen as to how the present favourable situation can best be turned to lasting advantage for the cause of freedom and world security.
>
> I hope you will permit me to say that I am afraid that, at the personal interview with which you honoured me, I may not have given you the right impression of the confidential conversation in Washington on which I reported. It was, indeed, far from my mind to venture any comment about the way in which the great joint enterprise has been so happily arranged by the statesmen; I wished rather to give expression to the profound conviction I have met everywhere on my journey that the hope for the future lies above all in the most brotherly friendship between the British Commonwealth and the United States.
>
> It was just this spirit of co-operation that I [came to England] in strictest confidence, to convey to you, that the President is deeply occupied in his own mind with the stupendous consequences of the project, in which he sees grave dangers, but also unique opportunities and that he hopes together with you to find ways of handling the situation to the greatest benefit of all mankind.

Oppenheimer acted quickly to maximise the laboratory's efforts to master implosion. Without either a plutonium gun bomb or implosion weapon, the burden would fall entirely on uranium and the less efficient gun method. Oppenheimer directed a major reorganisation of Los Alamos in July 1944 that prepared the way for the final development of an implosion bomb. Field tests showed that U-235 could be used in a gun-type nuclear device to meet the 1 August deadline but there was only going to be enough uranium for a single bomb so, unlike the plutonium bomb, there would be no opportunity for a test firing.

In the summer of 1944, however, it became clear that because of Pu-240 contamination, a gun-type design would not work for the plutonium bomb. The implosion method was now transformed from an intriguing possibility into a difficult necessity. Glenn Seaborg had warned that when Pu-239 was irradiated for a length of time it was likely to pick up an additional neutron, transforming it into Pu-240 and increasing the danger of pre-detonation. Measurements taken at Oak Ridge confirmed the presence of Pu-240 in the plutonium produced in their experimental pile X-10. On 17 July the difficult decision was made to cease work on the plutonium gun method and 'Thin Man' was abandoned. All effort was now centred on the 'Fat Man' plutonium implosion bomb. The first RaLa (radioactive lanthanum) tests were carried out under Louis Alvarez on 22 September in what has been described as the most important single experiment affecting the final bomb design. They were designed to study the behaviour of converging shock waves to achieve the spherical implosion necessary for compression of the plutonium at the bomb's centre. These tests were carried out in the Bayo Canyon outside Los Alamos to measure the collapse times and detonation and shock-wave velocities.

Abandonment of the plutonium gun project eliminated a shortcut to the bomb and necessitated a revision of the timetable Bush had given Roosevelt in 1943. On 7 August, Groves estimated that a small implosion weapons of uranium or plutonium could be ready in the second quarter of 1945 if experiments proved satisfactory and delivery of a uranium gun-type bomb by 1 August. It was anticipated that Germany might well have surrendered by the summer of 1945, making it likely that Japan would be the target of any atomic bombs ready by that time.

Three months later Bohr met Roosevelt who seemed more interested in some sort of approach to the Soviet Union although it is not clear what he meant by that. It was highly unlikely that he was about to give up any details of the Manhattan Project and was probably interested only in finding out what he could about the Soviet nuclear programme.

The Second Quebec Conference of September 1944 was held at the Chateau Frontenac Hotel and was the eleventh meeting between Churchill and Roosevelt during the Second World War. Roosevelt and Churchill signed a separate short agreement that became known as the Hyde Park Aide-Mémoire. Churchill had placed great store by his personal relationship with Roosevelt, telling Bohr on one occasion that 'there are [no problems] that cannot be amicably settled between me and my friend'.[19] Clause 2 of the document was hotly contested after Roosevelt's death and the very existence of the document, which had been signed in secret, would be questioned by the US until furnished with a copy by the British government.

> 1. The suggestion that the world should be informed regarding Tube Alloys, with a view to an international agreement regarding its control and use, is not accepted. The matter should continue to be regarded as of the utmost secrecy; but when a 'bomb' is finally available, it might perhaps, after mature consideration, be used against the Japanese, who should be warned that this bombardment will be repeated until they surrender.
>
> 2. Full collaboration between the United States and the British Government should continue after the defeat of Japan unless and until terminated by joint agreement.
>
> 3. Enquiries should be made regarding the activities of Professor Bohr and steps taken to ensure that he is responsible for no leakage of information particularly to the Russians.

Given how new all of the technological systems were, no one knew whether they would work as anticipated. Excitement mounted at Hanford as Fermi started inserting uranium into a reactor on 13 September 1944. There would be no going back: once the reactor was operational the intense radioactivity would make maintenance of many components impossible. From just after midnight until approximately 03:00 on 27 September, the pile ran without incident at a power level higher than any previous fission chain reaction but still at a fraction of design capacity. All looked good until, after three hours, suddenly the power level began falling and continued to fall until the pile closed itself down on the evening of the 28th. On the next day, mysteriously, the reaction began again, reached the previous day's level, then dropped. There was no way Fermi's team could explain what was happening.

Greenewalt stayed at the reactor until 02:00 working on solving the mystery and he found it to be on account of xenon, a fission by-product gradually building up and absorbing neutrons that should have been feeding the uranium chain reaction. When the pile shut down, the neutron flow resumed and the pile started up again. It was Greenewalt also who had insisted on the provision of a large number of extra tubes which now meant that the pile in B Reactor could be expanded to reach a power level sufficient to overwhelm the xenon poisoning. The first irradiated uranium was discharged from B Reactor on Christmas Day 1944 and, after several weeks of storage, went to the chemical separation and concentration facilities. By the end of January 1945, the highly purified plutonium underwent further concentration in the completed chemical isolation building, where remaining impurities were removed successfully. Los Alamos received its first plutonium from Hanford on 2 February.

On 17 December 1944 the 509th Composite Group under Lieutenant-Colonel Paul Tibbets was activated. This was a unit of the US Army Air Force equipped with modified B-29 Superfortress bombers, given the role of operational deployment of nuclear weapons and stationed in the Northern Marianas Islands in the Pacific Ocean. In particular, the overall weight of the aircraft was reduced to offset the heavy loads they would be required to carry. Rigorous recruitment was started for skilled machinists, welders and munitions workers with no more than a 20 per cent acceptance rate. By the end of May 1945 the unit would have an authorised strength of 225 officers and 1,542 enlisted men. Special equipment was designed to enable the unit to assemble atomic weapons at its operating base, thus allowing them to be transported more safely in their component parts.

In March 1944 Roosevelt's doctors revealed a variety of heart ailments and high blood pressure. For some time his appearance had become haggard and his lack of energy together with increasing lapses of concentration and memory were grave indications of rapidly failing health. His election victory over Thomas E. Dewey in 1944 and the Yalta Conference the following February had put the President under immense strain. In April 1945, he returned to Warm Springs, Georgia and on 12 April 1945, while sitting for a portrait, he collapsed and died of a cerebral haemorrhage. Vice President Harry Truman took the oath of office the same day.

A Target Committee was established on 27 April 1945, to determine the best techniques and targets in Japan to produce the most effective military destruction and psychological effects on the Japanese leadership. At the first meeting, sixteen areas in Japan were proposed for further study, of which four (Kyoto, Hiroshima, Yokohama and Kokura Arsenal) were ultimately chosen as potential targets for the bomb at the second meeting on 10 May.

As the US drew up plans to drop the first atomic bomb in 1945, a group of scientists at the University of Chicago, led by James Franck delivered a report arguing against the use of the bomb. The report stated:

> The development of nuclear power . . . creates grave political and economic problems for the future of this country.
>
> The scientific facts on which their construction is based are well known to scientists of other countries. Unless an effective international control of nuclear explosives is instituted, a race of nuclear armaments is certain to ensue following the first revelation of our possession of nuclear weapons to the world.
>
> We believe that these considerations make the use of nuclear bombs for an early, unannounced attack against Japan inadvisable. If the United States would be the first to release this new means of indiscriminate destruction upon mankind, she would sacrifice public support throughout the world, precipitate the race of armaments and prejudice the possibility of reaching an international agreement on the future control of such weapons.
>
> Much more favourable conditions for the eventual achievement of such an agreement could be created if nuclear bombs were first revealed to the world by a demonstration in an appropriately selected uninhabited area.
>
> If the government should decide in favour of an early demonstration of nuclear weapons it will then have the possibility to take into account the public opinion of this country and of the other nations before deciding whether these weapons should be used against Japan. In this way, other nations may assume a share of the responsibility for such a fateful decision.

The Interim Committee received the report on 12 June and reaffirmed its view that there was no alternative to the use of the bomb.

Chapter 9

SOVIET SPIES

'I did not have an uncritical view of the Soviet Union.'

Theodore 'Ted' Hall

A small group of scientists including Fuchs and Peierls had sailed from Britain to join the Manhattan Project on 3 December 1943 and landed from the SS *Andes* in Norfolk, Virginia. It is remarkable that Groves did not authorise a thorough security check on these British scientists given his almost paranoid suspicion of émigrés. He seems to have relied on MI6 to have done that for him but the British, despite running a third check, showed just how uncoordinated their security services were by not even realising that Fuchs had been given British citizenship. The British scientists set up office in 43 Exchange Place just a short walk away from the offices of the Kellex Corporation with whom they started working on the gaseous diffusion separation plant.

The British team had brought a more sophisticated design than the one chosen by the US but it was too late to change anything. Fuchs delivered his first report on 5 January 1944, 'Stability of Plant with Reset Controls', but a month later, he made a less conspicuous contribution to the evolution of atomic warfare. On a Saturday afternoon in Manhattan he met 'a short, full-faced man wearing a dark suit and an overcoat'. He wore gloves but also carried an additional pair. Carrying a book and a green tennis ball, Fuchs approached the man who asked him the way to Chinatown. 'I think Chinatown closes at five o'clock', Fuchs replied and with that he had made his first contact with his NKVD handler, whom he knew as Raymond, in the US.[1] It was an important meeting for the Soviets who had hit a brick wall trying to establish links to anyone of importance inside the US atomic programme. KGB records show that Fuchs was classified as 'a major figure with considerable opportunities and experience in agent work' but carried the caveat to the effect that 'caution when working with [Fuchs]' should be exercised. Raymond and Fuchs met monthly when Fuchs would pass handwritten documents covering British work on diffusion which confirmed that the US had put all its eggs in the gaseous diffusion basket. The reason for the handwritten copies was that all original copies bore identification marks and could be traced back to the person they had been distributed to, so any that might have been intercepted by G-2 would have meant disaster for Fuchs. Unfortunately, Raymond was not a very good handler. He was roundly criticised by Moscow for failing to run a thorough background check on Fuchs' personal situation. Had he done so he would have known that Fuchs lived alone in an apartment on the West Side rarely socialising and would have made arrangements to establish a back-up if he was to suddenly lose contact.

Raymond would turn out to be Harry Gold (Goose) who had grown up in a rough neighbourhood in South Philadelphia, where his family suffered discrimination as Jews and immigrants. During the 1930s he had been helped by Tom Black, a communist who tried to persuade Gold to join the Party but politics did not really interest Gold. The Soviet stance on outlawing anti-Semitism, however, did appeal to him and it was this that had allowed Black and his associates to draw Gold into the communist movement. A loner, Gold was a fantasist who liked to make up dramatic stories about a fictional existence involving himself and others which allowed him to slip comfortably into an espionage role, albeit a minor one. At first, he was given dreary and trivial spying tasks such as the copying of documents at his work, a sugar manufacturer. The drudgery of his work was unbearable for Gold who was bright and ambitious and saw a better life for himself which he hoped to find through studying part-time for a chemistry degree. This put him under a lot of time-pressure along with his job and frequent trips to New York as part of his espionage activity. Something had to give and he was on the verge of giving up his espionage work but was prevailed upon to stay by the New York *rezidentura* who offered to pay the tuition fees for a full-time degree course. Despite not being the best agent on the block, the Soviets had decided that he had long-term value as a courier but it was a mistake possibly brought on by the paucity of alternatives. Gold was, and never would be, well-grounded and he lacked the fundamental discipline for espionage work although he revelled in the novelty of it and loved demonstrating to Fuchs the tricks and ruses the KGB had taught him to avoid being followed in public places. Although his enthusiasm for spying waned after he graduated, he was persuaded to continue after the beginning of Operation Barbarossa when the enemy crystallised into the reviled Jew-hating Nazis and the US found itself on the same side as the Soviet Union.

Fuchs had been left in New York when Peierls had been sent to Los Alamos where they were developing a gun-type detonator for the uranium bomb but were finding great difficulty adapting it for plutonium. The war had taken a dramatic turn as Allied troops were now landing in Normandy but for Fuchs the future was becoming less clear. He was becoming disillusioned, with little to do in New York as his masters considered whether he would return to Britain to work on the design of a gaseous diffusion plant or go to Camp Y in Los Angeles. At the end of July, the decision was made for him to return to Britain, but when Oppenheimer heard about it he was not prepared to let such a valuable asset go and arranged for Fuchs to go to Los Alamos instead and the transfer was made immediately. At his next scheduled rendezvous with Fuchs on 5 August Gold carried $500 he had been authorised to give to Fuchs as payment for his espionage work, but he waited near the Brooklyn Museum of Art rendezvous in vain. Fuchs failed to show and neither did he turn up to the next scheduled meeting. Gold's poor tradecraft was coming home to roost.

Moscow was not impressed with the way Gold was handling Fuchs. A detailed complaint was filed against him citing his failure to establish almost anything at all about Fuchs' personal life and particularly the lack of any information that would now help him to track Fuchs down. Fortunately for him, Gold was given Fuchs' home address, 128 West 77th Street. It was yet more poor tradecraft, but he had little choice now but advertise their connection by going there to look for him. Getting no response at the door, Gold was told by a neighbour that he believed Fuchs had left the city in something of a hurry. Clearly he had left before he was able to meet Gold at the next scheduled meeting and he had no other way of making contact. 'Contact lost' was the message that went back to Moscow and Gold waited nervously for the repercussions.

In Los Alamos Fuchs was once more working under Peierls, this time on calculations for and design of the implosion mechanism that was the triggering device for the plutonium bomb. Explosive material surrounding the plutonium core had to be detonated powerfully, symmetrically and simultaneously to crush the core to start a chain reaction. The firing mechanism, shape and speed had to be exact down to a tolerance of one-millionth of a second. Fuchs made or reviewed the theoretical calculations supporting many of the components in the mechanism.

In the meantime, Gold had been told where Fuchs' sister, Christel Fuchs Heinemann, lived and so on 24 October he knocked on the door of 144 Lakeview Avenue in Cambridge, Massachusetts. Fuchs had made arrangements with Christel in case he had to leave in a hurry, so he had primed her to respond positively if someone made contact and identified themselves with a code word. Gold did this but Christel could only tell him that Fuchs had gone away, probably back to Britain. With little else to go on, Gold returned a week later hoping that Fuchs might have got in touch and sure enough, Fuchs had made a trip to Chicago and called his sister to say that he was now working in New Mexico and would visit her for two weeks at Christmas. A month later, Gold returned again to Christel who told him that Fuchs might have to go to New York for a few days so Gold gave her a telephone number that Fuchs could call if he wanted to see him. Fuchs didn't make it for Christmas but did get to Cambridge in February when Christel gave him Gold's number. The two men met up and Fuchs gave Gold details about the principles of the plutonium bomb, its construction, dimensions and the trigger, types of explosives and their timed sequencing, They then made arrangements to meet in Santa Fe after Gold had relocated to the West Coast. On 2 June 1945 Fuchs went into Santa Fe ostensibly to shop but it was the day for his meeting with Gold arranged a few months before in Cambridge. Fuchs brought with him more plans for the plutonium bomb that was to be tested on 16 July. He was at the centre of research and was able to provide a sketch of the bomb components, an estimate of critical mass as well as dimensions and details of the implosion method of detonation. This was described by Igor Kurchatov as 'exceptionally important' in as much as it ruled out the possibility of using the gun method to detonate plutonium. It was clear that Fuchs was 'by far the most important informant in the Manhattan Project'.[2] Upon his return to Britain, Fuchs was appointed Head of Theoretical Physics at Harwell but before he left the US he had been involved in discussions about a new thermonuclear bomb and, even though Igor Gouzenko's defection had damaged the Soviet espionage networks in the US and Canada, it is possible that Fuchs' notes on this found their way back to Moscow via Lona Cohen.

The former KGB officer turned journalist Alexander Vassiliev was given the unique opportunity to spend two years poring over materials from the KGB archive in 1993, taking detailed notes particularly about Soviet espionage in the United States during the years 1930–50. One of the first important revelations from Vassiliev's notebook confirms what was later known that Julius and Ethel Rosenberg, who were executed in 1953 by the US government for the so-called theft of the atomic secrets, were innocent of that crime. Julius, 'an earnest and ideologically committed, if rather bumbling, spy', was certainly a Soviet agent but despite FBI Director J. Edgar Hoover famously calling his offence 'the crime of the century', the secrets he stole were more to do with radar, fuses and electronics. Whilst enthusiastic about his espionage work, Rosenberg was impetuous and clumsy to that extent that he fumbled and failed with the Minox camera he used when photographing secret documents and had to be continually instructed on its use by his handler Alexandr Feklisov aka Aleksandr Fomin. It was Ethel's brother, David Greenglass (Caliber), a machinist at Los Alamos, who did supply nuclear weapons secrets to the Soviet Union.[3]

The United States government routinely acquired copies of all cables openly sent to and from various Soviet embassies and consulates (code name Venona decrypts) but these messages were encrypted and the method used 'the one-time pad' should have made them impossible to decipher but the US Army's Signal Intelligence Service gradually discovered a Soviet procedural error that allowed many of the archived messages to be read after 1946. These messages contained references to spies who had been active throughout the war and had never been detected, including many who presumably were still working for Soviet intelligence. Under the NKVD code name 'Rest', Fuchs was identified by the Venona decrypts and arrested in Britain in January 1950. He was tried, was found guilty of espionage and served a little over nine years of a fourteen-year sentence. Upon his early release for good behaviour, Fuchs moved to East Germany, where he headed a scientific research institute outside Dresden. He died in 1988 at the age of 79. Vassiliev's notebooks confirm that Fuchs had in fact continued to spy until at least 1949. Fuchs' undoubtedly major contribution to the Soviet atomic bomb project efforts were recognised and efforts made to reward him with honorary membership in the Soviet Academy of Sciences but Soviet scientists objected, fearing that it would play down their own contribution to the development of the Soviet atomic bomb. Instead, in 1989 the KGB would persuade the Communist Party's Central Committee to grant Fuchs posthumously its Order of Friendship among Nations.

In mid-1944, 30-year-old chemist Martin Kamen, a rising star at Berkeley's Radiation Laboratory, was summarily fired from the Manhattan Project by Groves. On 1 July 1944, he had been photographed arm-in-arm with two Soviet spies, Grigory Kheifets (Charon) and Grigory Kasparov (Gift), at a San Francisco restaurant. Having been tipped off by a 'talent-spotter', the English physicist George Eltenton (Dorin), Kheifets was sounding Kamen out as a possible recruit.

In a roundabout way of inveigling him into the first stages of a covert relationship, he told Kamen that they were interested in obtaining medical treatment for an ailing colleague in Portland and Kamen had given them unclassified scientific articles on the therapeutic use of radioisotopes for cancer sufferers. Unfortunately for Kamen, Kheifets, who was the San Francisco Soviet station chief, was being tracked by both G-2 and the FBI who had been listening in to their dinner conversation with hidden microphones.

Subsequent to his firing, Kamen was blacklisted by the US government and unable to obtain a passport for the next decade. At one point he attempted suicide as a result of being continually hounded by the FBI, Congressional investigators and journalists who portrayed him in lurid newspaper accounts as a Soviet spy. Although a promising scientific career had been ruined, Kamen's personal reputation was finally rehabilitated and his innocence confirmed when Vassiliev's notebooks were published in 1996 after which he was awarded the prestigious Enrico Fermi Prize by the US Department of Energy. Eltenton had worked for some time in Leningrad before moving to Emeryville close to Berkeley where he became active in union recruitment. He had also been complicit in an unsuccessful attempt to persuade Oppenheimer to pass secrets to the Soviets back in 1943 because he had openly declared that he 'did not feel [the Soviet Union] was getting the scientific and technical support it deserved'.[4] After interviews with the FBI and facing a call to appear before the House Committee in Un-American Activities (HUAC), Eltenton fled the US in 1947.

Both the GRU and the NKVD had been unsuccessfully trying to recruit Oppenheimer since January 1943 and in 1944, Eltenton and Kheifets had positively identified Oppenheimer as an important member of the Manhattan Project and given him the code name 'Chemist' while Oppenheimer's brother, Frank, was called 'Ray' although their reports show that Oppenheimer's role as director of the Manhattan Project was not known to them. He was sufficiently important, however, to be 'of great interest' to the NKVD who, given Oppenheimer's well-known leftist leanings, had 'reason to expect that his cultivation would yield positive results'.[5]

Louise Bransten was another of Kheifets' contacts. A 33-year-old San Francisco socialite, known locally as the 'Apricot Heiress', she had inherited a fortune from her parents' dried fruit importing business. In a November 1944 assessment of Kheifets' work, his immediate superior, Pavel Fitin (Viktor), the wartime head of the NKVD's foreign intelligence directorate, wrote despairingly that 'despite Louise

Bransten's extensive connections in various US political and financial circles, [Map] has not given us anything of value'. This came on top of a recent Moscow assessment that the New York *rezidentura*, despite swallowing many millions of dollars, was recruiting useful agents at a pace that was 'particularly intolerable'.[6] New York parried the blow by citing Fuchs as a major catch but Fitin was quick to point out that it was not they who had recruited him. Kheifets was equally dismissed as having 'failed to organise himself being obsessed with quantity and not quality'.[7] He had failed to provide any usable contact for two years when he was recalled to Moscow for 'failing to cope with his job'. He was replaced by Kasparov who remained in place for only a few months before being transferred to Mexico City. Kheifets was facing an ignominious return to Moscow having already been dismissed from the intelligence services in 1938 but subsequently reinstated. It is clear from Soviet archives that agents in the US at the time were repeatedly overstating the extent to which they had penetrated the Manhattan Project and often conspired one with another to support each other's claims.

Artur Alexandrovich Adams was born in Sweden and taken to Russia by his mother before being orphaned at age 10. In St. Petersburg he was converted to Marxism and suffered police brutality and internal exile for his part in the 1905 Russian revolution. Later as a student he travelled widely to Egypt, Canada, Italy and Argentina as well as the USA to which he returned on a number of occasions. In 1938 he entered illegally through Buffalo posing as a Canadian radio engineer. He set up a business, Technological Laboratories, which gave him ample excuse to travel across the whole country. It was during his travels that he first became aware of the Manhattan Project through one of his contacts Clarence Hiskey (Aesculapus). Through Hiskey, Adams met John H. Chapin who had worked at DuPont before joining the Manhattan Project. Chapin proved to be a particularly productive source especially in relation to plutonium and the enormous level of US investment in its production. This was a short-lived victory, however, because when Adams tried to give Chapin some money for the information he had produced, it unnerved Chapin to the point that he informed the FBI. He obviously felt that it was not a crime to pass intelligence to an ally in wartime but being a paid informer was another thing entirely. The FBI set up surveillance on Adams and when this was picked up by the GRU, he was put into cold storage as an agent and attempts were made to get him out of the country first through Portland, which failed, and then after a year hiding out in New York he eventually made it back to the Soviet Union.

January 1943 saw the arrival in New York of Leonid Kvasnikov, ostensibly as part of the large contingent of Soviet personnel involved with the Lend-Lease (Amtorg) programme of US arms going to the Soviet Union, the country which had now been temporarily replaced as the repository of all evil by Nazi Germany. It was an opportunity that the Soviets were not likely to overlook to swamp the US with intelligence agents and spies of all stripes. When Kvasnikov's man, Gregori Kheifets, started trawling for contacts it was what the military men would call a 'target-rich environment'. It was Groves' worst nightmare. Some like Fermi and Pontecorvo had been approached back in the early 1930s in Rome because of their anti-fascist views and it was no problem for Kheifets to strike up a new conversation with them but it was Oppenheimer who was seen as the main target. Somewhat crude and unsuccessful attempts were made to sound him out. The US security services attached to the Manhattan Project spent a lot of time marking their turf and vying for dominance but they soon found plenty of more important things to be worried about. Surveillance on Steve Nelson, a key communist activist in San Francisco, caught him making contact with Joseph Weinberg, one of a number of members of Oppenheimer's team who were known to have communist sympathies.

Steve Nelson had been born Stephan Mesaroch in Chaglich, Croatia and had arrived in the US illegally in 1920 using the name Josef Fleischinger. The deception was uncovered and deportation procedures were initiated. For some reason these were abandoned two years later and he was allowed to stay with his mother and two sisters who had come over with him and they were all given US citizenship in 1928. Later he spent some time at the International Lenin School in Moscow before travelling to India and China. After returning to the US and becoming involved in organising unions in Pittsburgh, Chicago and Cleveland, he served in the Spanish Civil War as political commissar in the Abraham Lincoln Battalion. It was while he was in Spain that he acquired another US passport in the name of Josef Fleischinger and it was probably here also that he was recruited by the NKVD.

He ended up in San Francisco as part of the Soviet Enormoz spying operation and was run as an agent by Vasili Zubilin who was part of Kvasnikov's New York *rezidentura*. Nelson was the 'bag man' who handled all payments to the extensive spy ring that the Soviets were running all across the USA Nelson's apartment was bugged by the FBI and revealed that he had handled highly secret information from such as Weinberg which was passed on to Ivan Ivanov of the Soviet Consulate. There is convincing evidence that Nelson approached Oppenheimer with a view to recruitment, but he was flatly turned down.[8]

Zubilin would actually be later accused of misappropriating $80,000 of the Soviet 'slush fund'.[9] The FBI had not been informed about Los Alamos or the Manhattan Project as a whole and were slow to become aware of the Soviet spy network. Nelson would, however, later be convicted in July 1952 of 'conspiring to unlawfully, wilfully and knowingly advocate and teach the duty and necessity of overthrowing and destroying the Government of the United States by force and violence, with the intent of causing the aforesaid overthrow and destruction of the Government of the United States by force and violence as speedily as circumstances would permit'. Sentenced to 20 years in jail, he was released on appeal three years later.

Groves was fully cognizant of Oppenheimer's background and relationships, especially with Jean Tatlock, a prominent Communist Party activist. Tatlock was very beautiful with 'long dark hair, red lips and hazel blue eyes beneath heavy lashes' but she could be 'moody and introspective', suffering severe bouts of manic depression.[10] She and Oppenheimer had been very close and had almost married in the late 1930s but separated before Oppenheimer married Katherine Peuning on 1 November 1940. Notwithstanding their estrangement, G-2 security agents reported that Oppenheimer and Tatlock met again secretly in San Francisco in 1943. In a 1954 US Department of Energy security board hearing, Oppenheimer called her 'not much of a communist' and admitted that when they met again they still had 'very deep feelings' for each other. This had been a very troubling development for Groves.

Tatlock was found dead on 4 January 1944 by her father who broke into her apartment after failing to make contact with her. He found her lying on a pile of pillows at the end of the bathtub, with her head submerged in the partly-filled tub. A suicide note was found close by. Speculation surrounding Tatlock's death focused on the figure of Boris Theodore Pash, the head of G-2. Pash, born Pashkovsky in San Francisco, claimed to have remembered the earthquake of 1906 before going to Russia with his father who was a Russian Orthodox bishop. He fought for the counter-revolutionary White Army in Odessa where he took a bayonet wound in the leg. Forced to flee from the Crimea as Bolshevik forces took control, he narrowly avoided a massacre and forever nurtured a pathological hatred of communism. When he returned to the US with his wife and son in 1924, he shortened and anglicised his name to Pash. He was regarded by fellow Russian émigrés as 'a really wild Russian, an extreme right wing, sort of Ku Klux Klan enthusiast'.[11] As head of G-2 he had attempted to get Oppenheimer fired as a potential spy because even if Oppenheimer wasn't himself spying, he might be setting up people within his organisation, with Tatlock as the conduit, ,but he was overruled by Groves. In the mid-1970s, during the Church Committee hearings about the CIA, it emerged that from 1949 to 1952, Pash was Chief of Programme Branch 7, which was responsible for assassinations, kidnappings and other 'special operations', but there is no evidence to suggest that he ever carried out any killings personally.

In 1942, Pash had been a counterintelligence officer in the Western Defence Command and had established an elaborate system of investigation into communist espionage activities. By early 1943 he was carrying out complex investigation of communist activity within the Manhattan Project. His team followed the suspected Soviet agents and installed microphones in their houses. He had delayed giving Oppenheimer security clearance for a full three months after Groves had selected him to run Los Alamos. Pash felt that not only should Oppenheimer be denied clearance, but he should also be taken off the project completely because of his 'support for every left-wing organization on the West Coast'. Groves discussed the matter with Lansdale, his special assistant for security who coordinated the activities of the Manhattan Project's field security teams with those of other agencies such as the FBI. Both men were confident that Oppenheimer was not a security risk and Groves gave Pash a direct order to expedite his security clearance.[12]

Pash was quite correct to be extra cautious about security. The full extent of the Soviet spy ring would become evident only much later in September 1945 when Igor Gouzenko, a cipher clerk at the Soviet Embassy in Ottawa on the staff of Colonel Nikolai Zabotin, gave himself up to the Royal Canadian Mounted Police (RCMP). Gouzenko, described as 'a man of humble Russian origins and modest height', had decided to defect when orders came through for him and his family to return to the Soviet Union. The prospect of leaving a thriving Western city to return to drab war-torn Moscow was not appealing but neither was there any certainty of a friendly welcome and quiet resettlement in Canada if he defected. He was, however, prepared to take that chance.

On the evening of 5 September Gouzenko, carrying a briefcase stuffed with Soviet code books, entered the offices of the *Ottawa Journal* and blurted out a stuttering, largely incomprehensible warning in broken English about the Soviet Union after which a puzzled sub-editor advised him to contact the police but when he went to the offices of the Justice Ministry in Wellington Street the late hour meant that nobody could see him until the following day. Gouzenko was now in no-man's-land, unable to find the security of the Canadian authorities and burdened with the fear that his planned defection and treason would soon be discovered by Zabotin.

The next morning, Gouzenko returned to the Justice Ministry with his wife and child and again with his incriminating briefcase, explaining that he wanted to defect and that he had important information about a Soviet spy ring operating in Canada. He tried to impress upon everyone the danger that he and his family were now in but he was denied an interview and neither would the staff of the *Ottawa Journal* take him seriously. He had not been ignored, however: it was just that nobody seemed to know what to make of him and nobody was willing to be responsible for precipitating a diplomatic incident. Word had quickly gone up as far as the desk of the Canadian Premier Mackenzie King who did not want anything to do with Gouzenko and the advice that came back down the chain was that he should go home and return the stolen documents.

Nothing daunted, the Gouzenkos then went to the Canadian Crown Attorney's office to apply for naturalisation only to be told that the process could take several months. They returned to their home at 511 Somerset Street in a state of utter despair. It seemed only a matter of time before Gouzenko's theft of the documents was discovered and so it was not a total surprise when that evening there was a loud banging on his apartment door by someone who identified himself as a member of the Soviet Embassy. Gouzenko went out onto his balcony and managed to contact his neighbour, a Royal Canadian Air Force corporal, Harold Main. Main heard the knocking and agreed to go on his bicycle to get the police. After he left, the banging stopped and Gouzenko ventured out into the hall with his heavily-pregnant wife Svetlana and two-year-old son Andrei. Another neighbour, Frances Elliot, listened to their story and bravely agreed to take them in.

Main returned with police officers, but with nothing to see they showed little interest. Without knowing the full story of Gouzenko's attempted defection, which he was unwilling to divulge, the officers explained that they were unable to do anything, since the apartment was considered Soviet property and they soon left. Before too long four members of the Soviet Embassy staff returned and battered down the front door of Gouzenko's empty apartment. They emptied drawers and searched the rooms for the missing code books. Main once again sped off to inform the police who returned this time to find the intruders ransacking the apartment. It was still a delicate diplomatic issue for the police to deal with but they were able to persuade the intruders to give their names and ranks before getting them to reluctantly vacate the apartment.

The next morning, the Gouzenkos were taken a little more seriously and were escorted to a cottage outside Ottawa where they were interviewed by the Canadian security services. They were granted asylum and the RCMP staged a mock search for them after being alerted by the Soviet embassy about their disappearance. Gouzenko's stolen documents subsequently revealed the huge scale of Soviet penetration of Canadian agencies. A report of the Royal Commission in Ottawa in 1946 concluded that there had been two parallel Soviet spy rings in Canada, a Soviet military espionage network run by Zabotin in Ottawa and the other an NKVD network run by Vitali Pavlov, and there seemed to be no love lost between the two of them. Of Gouzenko, the Commission reported that:

> He has undoubtedly been a most informative witness and has revealed to us the existence of a conspiratorial organization operating in Canada and other countries. He has not only told us the names and cover names of the organizers, the names of many of the Canadians who were caught 'in the net' (to employ the phrase used by the documents) and who acted here as agents, but he has also exposed much of the set-up of the organization as well as its aims and methods here and abroad.

He was given a new identity as George Brown and his family spent the rest of their lives in Toronto.

One of the spies unmasked by Gouzenko's defection was a 'retiring and lonely man' called Alan Nunn May, a Cambridge physics graduate who had been a Communist Party member since the early 1930s and had made a number of visits to Moscow before the war. Recruited to Tube Alloys while working at the Cavendish, he later went with the British team of scientists to Canada in 1943 where he made contact with Zabotin. Over the next two years Nunn May's work took him to several of the Manhattan Project sites including the Argonne Laboratory in Chicago. He initially had no scruples about passing information to the Soviets about his own research which included US intelligence about atomic research in Germany but he later claimed that he had never benefited financially from his treason. He would later serve six years in Wakefield prison before going to Ghana as a government adviser on scientific matters.

Sergei N. Kurnakov (Bek) had emigrated to the US soon after the Bolshevik Revolution by way of Yugoslavia and France. In New York he worked for Soviet Intelligence under the cover of being a journalist reporting on military issues for the *Daily Worker* and *New Masses*. On 12 November 1944, Kurnakov sent a report to Moscow in which he described a 19-year-old graduate of Harvard as both a talented physicist and member of the Young Communist League. This was Theodore 'Ted' Hall (Young) currently working at Los Alamos who had just made contact with Kurnakov and handed him a dossier about his place of work. Hall, a 'gifted and moody teenager',[13] was the son of a Russian furrier, who had entered Harvard at age 16 to study physics and mathematics and been influenced by a fellow student Saville Sax who understood Hall's deep anxiety over the US becoming the only nuclear power in the world, but Hall would later say that he had never been actually recruited to the NKVD, having volunteered to work for them.

Initially Hall and Sax, through the Communist party newspaper the *Daily Worker*, approached Kurnakov who at first suspected an entrapment ploy and took some convincing even though he had already received a handwritten report on the work Hall was doing. Eventually Moscow felt that it was worth the risk and Kurnakov was instructed to make an encouraging reply but by this time Hall was on the point of leaving New York. Kurnakov was forced to make contact under rather embarrassing circumstances as Hall stood on Penn Station with his parents as he prepared to return to Los Alamos. It was agreed that Sax would be Hall's go-between and they worked out a rather crude communication code for arranging clandestine meetings based on Walt Whitman's *Leaves of Grass*. Whitman had attached to each of his verses a number from one to thirty-one that was ideal for days of the month. Much to Kurnakov's chagrin, Hall was quickly transferred to be run by Anatoli A. Yatskov, a parvenu. Looking at the sort of material Hall was providing it is no wonder that Kurnakov was aggrieved. An agent who could provide information about not only all the working plants in the Manhattan Project but descriptions of all four methods of uranium purification was going to put his handler in an excellent position for promotion. Furthermore, he had a much better grasp of nuclear physics than Yatskov who was soon berated by Moscow for his poor performance. It was Hall, however, who had initiated the switch of handler, having become irritated by Kurnakov's poor security procedures.

Any lapse on Kurnakov's part, however, was dwarfed by the ineptitude of Hall's own transgressions. A meeting arranged between Hall and Sax in Albuquerque had them both moving towards the rendezvous on foot, early and from the same direction which caused them to bump into each other in the street. They quickly moved to a quiet spot where Sax gave Hall a piece of paper he had brought from New York inside his shoe requesting information about sulphur dioxide which was an ingredient used to calibrate the critical mass of uranium. In return, Hall gave Sax a handwritten description of a method for assembling a critical mass that proved to be a crucial piece of information leading to the successful production of a plutonium bomb.[14]

Chapter 10

GROUSE, FRESHMAN AND GUNNERSIDE

'ingen adgang unntatt på forretningsreise
kein Zutritt außer geschäftlich
[No admittance except on business]
Sign on the door of the heavy water room at Norsk Hydro

All intelligence coming out of Germany in 1941 had indicated that Heisenberg was leading a concerted effort to build a German atomic bomb in a process that required large quantities of heavy water but that he had, so far, not been able to establish a chain reaction. But the fear, as reported by Compton in Washington, was that even if they did not have an actual atomic bomb 'there is a real danger of bombardment by the Germans . . . using a bomb designed to spread radioactive material in lethal quantities'.[1]

In February, British Intelligence had already started making contingency plans to sabotage the Swedish port of Oxelosund to disrupt the supply of iron ore to German industry and had also set up a network of agents in Norway who, amongst other things, maintained surveillance on the Vemork fertiliser plant. The MI6 officer in charge of Scandinavian operations was Lieutenant-Commander Welsh, a man who 'took an exceptionally personal approach' and was very highly thought of by all his agents who styled themselves 'Welsh's gang'. Welsh had learned from Leif Tronstad, a pre-war professor of chemistry and a member of the British network in Norway, about the Vemork heavy water plant and about German efforts to harness atomic energy.[2]

In May 1941, Karl Wirtz at the Uranverein came up with a new catalytic method to speed up the production of heavy water and went to Norsk Hydro to see about incorporating the method into their production process. News of this leaked out to Norwegian Resistance groups that had been instructed to find out what was going on at the Vemork plant and where in Germany the heavy water was being sent. Einar Skinnarland had joined one of the small group of Norwegian Resistance fighters (Norwegian Independence Company No. 1) that had been formed by the British Special Operations Executive (SOE) after the German occupation. He was brought to Britain and briefed by Tronstad, who had fled Norway and was now head of Section IV of the Norwegian High Command operating in Britain with the SOE. After a crash-course in espionage in the Scottish Highlands, Skinnarland was parachuted back into Norway close to Rjukan in March 1942.

Jomar Brun had been head of hydrogen research at Norsk Hydro since 1933 when heavy water was first produced. The first consignment that left the plant in August 1934 had gone to Birkbeck College in London and Brun and Tronstad had published a paper on the physical properties of heavy water in *Nature* in 1935. Brun was also friends with Karl Wirtz and the two had corresponded with each other over the years. Tronstad now got in touch with Brun, asking him to keep Britain informed about developments in the German effort to increase the production of heavy water which he referred to as 'juice' and to see if there was any way that some of the 'juice' could find its way to British research labs. No practical way of doing this could be found but Brun did take it upon himself to reduce supplies to Germany by sabotaging the production of heavy water. He introduced quantities of castor oil into the electrolyte used for extraction and this, along with other small acts of sabotage, had the effect of closing down the plant for a time in April 1942. Obviously this level of sabotage would not have a decisive effect on heavy water production and would be uncovered if it persisted for long so discussions were held between SOE and the Norwegian Government-in-Exile to see what more could be done to exploit the 'Achilles heel' of German nuclear research.

The long days of the Norwegian summer left few hours of darkness for mounting operations against the plant, but Churchill had designated Vemork as a 'top priority target'[3] and so Brun made micro-copies of photographs and plans of the plant that were smuggled out in tubes of toothpaste through Sweden to Britain. The British War Cabinet requested an urgent review of what options were available to close down the plant. The disappointing response was that there were a number of possible avenues but they all had in common the fact that they were extremely risky and of doubtful outcome. Bombing from the air risked civilian casualties while a commando raid was thought unlikely to succeed at best and suicidal at worst, with inevitable reprisals against the local population.

Nevertheless, a plan, code-named Grouse, was hatched for dropping a raiding party of ten commandos by parachute onto the frozen wastes of the Hardanger Plateau. They would then reconnoitre and identify a site where a second larger party would come in by glider. After destroying the plant, the men would trek across 250 miles of mountains to Sweden. Wise heads exposed the many probably deadly flaws in the plan, but it was given the go-ahead.

After a couple of attempts that had to be aborted, the much-reduced four-man Grouse team led by Second Lieutenant Jens Anton Poulsson (cover name Jens Dale), including Claus Helberg (Sverre Haugen), Knut Haugland (Knut Strand) and Arne Kjelstrup (Arne Kristoffersen), dropped onto the Hardanger Plateau on 18 October 1942, 30 miles from the plant, the aircraft navigator being confused by the barren landscape's lack of navigable clues. Almost immediately storms erupted and, as the long nights set in, the party took 15 days to plough through fresh snow and cross fast-flowing mountain streams. Each man had to make the journey twice to transport the great deal of equipment they had brought with them before making contact with Skinnarland, who had been back in Norway since 28 March. On 9 November, Poulsson reported 'I fell through the ice, whilst crossing a river . . . it was the second time'. When Poulsson signalled that a 'lake covered with ice and partly covered with snow' some three miles from the plant was suitable for the gliders to land the second part of the plan, Operation Freshman, was given the green light by Churchill.

Two Halifax bombers set off from Skitten airfield on the northern tip of Scotland on the night of 19 November to make the 800-mile round trip. Behind them each towed an Airspeed Horsa glider with seventeen volunteers from the Royal Engineers of the 1st Airborne Division. One glider was piloted by Sergeant M.F.C. Strathdee and Sergeant P. Doig of the Glider Pilot Regiment and the other by Pilot Officer Davis and Sergeant Fraser of the Royal Australian Air Force. The thirty sappers were led by Lieutenant G.M. Methven. Nothing like it had ever been attempted before and there had been little time to train the crews for the testing navigation and weather conditions they would encounter. It was estimated that a party of fifteen was sufficient to complete the mission but two parties were send as a precaution.

The first Halifax and glider combination took off at 17:50 and encountered poor weather conditions. Things started to go wrong right from the start. Almost as soon as the aircraft was in the air, the communication system connecting the glider to the mother craft failed. That was not enough to abort the mission so the Halifax ploughed on with the commandos being dragged through the bitter night air across an icy North Sea deep into hostile territory. After hours of darkness the aircraft searched in vain for the dropping point having failed to locate the homing signal from 'Eureka' beacons set up by the Grouse party. The navigator was left with no option but to plot his route as best he could across a wilderness of almost complete featureless white with no observable markers to go by.

The pilot, Squadron Leader A.B. Wilkinson and the onboard mission commander, Group Captain T.B. Cooper DFC, could not fix their position. Running low on fuel and rather than cut loose the glider without knowing where they were, Wilkinson turned for home taking the glider back with him. They made it as far as the Norwegian coastline before the towline froze and as the next rush of turbulence jostled both craft, it snapped like a glass rod setting the glider, Horsa DP349, free-falling to earth and crashing into a mountain near Fyljesdalen. Cooper mistakenly thought the glider may have gone down into the sea and vainly called to base for a search and rescue mission if any British vessels were in the area. There were none. Eight men on board DP349 were killed as it plunged into the mountainside. The survivors, four of whom were badly injured, were quickly rounded up by local German troops and interrogated. The injured were tortured, three of them being strangled half to death with a leather strap before being killed by having air injecting into their bloodstream. The fourth was executed by a gunshot to the back of the head. The other five were incarcerated in the Grini concentration camp near Oslo before being murdered by firing squad.

Halifax W7801, piloted by Flight Lieutenant Roland Parkinson, RCAF, took off at 18:10 and went in at sea-level to avoid radar but got into trouble as it reached the Norwegian coast when it tried to gain altitude. It smashed into a mountainside, setting free the glider which came down heavily close by at Helleland some 155 miles from Vemork. All seven crew of the Halifax were killed in the crash. Two men on board the glider were killed instantly and a third injured. The survivors were quickly rounded up by German soldiers, interrogated and summarily executed.

In a piece of stomach-turning irony, a Hitler Directive of 18 October had authorised the execution of commandos and special forces since their deployment 'did not conform to the principles laid down in the

Geneva Convention' and 'all quarter is to be denied them on principle'. As if Hitler cared about the Geneva Convention or principles! Security around the Vemork plant was tightened by minefields laid all around the perimeter and the local Wermacht garrison was reinforced. Einar Skinnarland managed to evade capture but his brothers, Torstein and Olav, were rounded up and thrown into Grini concentration camp. In September 2004, a team conducted a successful expedition to Norway to recover the wreckage of Horsa DP349.

On 20 November 1942, the following announcement was made over German radio 'On the night of November 19-20th two British bombers, each towing one glider, penetrated Southern Norway. One bomber and both gliders were forced to land. The sabotage troops in the gliders were engaged and killed to the last man.'

Grouse now became Swallow and Poulsson's small band of commandos settled in to survive the bitter Norwegian winter. Their food supply was soon exhausted and they hunted reindeer to survive. Poulsson reported that they lived entirely on meat, eating 'every morsel of the animal in Indian fashion [and] eating the stomach with all its contents for vitamin C' which he called 'a delicacy'.

A second attempt, Operation Gunnerside, was made to sabotage the plant on 16 February 1943 when a team of volunteers led by Lieutenant Joachim Rønneberg (cover name Jens Haugen) included Knut Haukelid (Knut Hartmann), Kasper Idland (Karl Olsen), Hans Storhaug (Hans Kylling], Frederik Kayser (Fredrick Fjell) and Birger Strømsheim (Birger Sunde) was parachuted into Bjornesfjorden, about 30 kilometres from Swallow's camp.

Rønneberg later recounted how difficult it had been to prepare for the trip, especially getting clothing to keep out the bitter winter weather. Maps of southern Norway showing where German troops were stationed were provided by the Norwegian Resistance. The first attempt to enter Norway in January was aborted when the aircraft failed to make contact with the Swallow team but they tried again on 17 February when they parachuted in but landed 15½ miles from the rendezvous spot. They eventually met up with Swallow on 23 February, then on 25 February they moved out, headed for the Norsk Hydro plant. Intelligence from the Norwegian Resistance was comprehensive and indicated that 'there was a German barracks at the plant containing some 20 soldiers . . . a double guard down on the bridge . . . a gate in the middle of the bridge and a sentry box with a button to set off the alarm to the guards at the factory. Everything was fenced in, there were mines all along the water pipeline, except for the entrance to the railway which was on the southern side of the valley.'[4]

It was not possible to move on the plant from the heights due to the depth of snow and they were told that crossing the ravine was impossible. The Norwegian Resistance suggested they mingle with workers going into the plant for their shift but the certainty of reprisals to the local community ruled that out. The saboteurs were all wearing British uniforms to make it clear that if they were taken or their bodies found the local population would be seen not to have been involved. Eventually a crossing point of the ravine was located from north to south some hundreds of metres below the bridging point. 'We left our equipment near the road leading down to Rjukan and had only the explosives and the weapons with us when we crossed the river', Rønneberg recalled.[5] Just after midnight they cut the barbed wire and the covering party took up positions while the demolition party entered the building. They started laying the four pre-formatted 4.5kg charges with two-minute fuses which were shortened to 30 seconds. They told a Norwegian worker who stumbled upon them to go into another room and lie down with his mouth open. He must have done as he was told because years later he met some of the saboteurs again and he was not deaf which he would have been if his mouth had been closed when the charges went off.

The men quickly returned the way they had come as the charges exploded and set out on a three-hour climb up the 700m mountainside through thick snow on the northern side of the valley. By now the alarm had gone off and there was a lot of traffic headed for the plant but high up on the plateau, the saboteurs put on their skis and left as quickly as possible. The German guards knew that nobody had crossed the bridge and, thinking that the ravine was impassable, started searching on the south side of the valley. The next day there was a storm with fresh snow to cover any tracks. The men returned to Skryken where they had left their equipment with Swallow. Five of the party crossed into Sweden on 18 March claiming to be refugees. In Stockholm, Rønneberg and some others of the mission crew went to the opera to see *La Traviata* before returning to London on a mail aircraft while the others remained behind, including the wireless operator who kept transmitting. All the members of the Gunnerside mission survived the war.

It had been estimated that all the vital parts of the plant machinery were blown to pieces and 500kg of heavy water was destroyed, along with all the equipment used to produce it, but what damage had been done was quickly repaired and the production of heavy water started up again within weeks. Skinnarland radioed to London that full production would start in August. SOE still considered it vital that heavy water supplies to Germany were stopped but another commando raid was out of the question. A straightforward aerial bombing attack seemed the only way but the British Chief of the Air Staff was not enthusiastic given the difficulty of hitting the target so tightly set against the mountainside and the chances of placing bombs onto the critical section were remote.

The US Air Force, however, did not agree. Groves asked OSS to find out exactly what the situation was at Vemork. Donovan chose to send Moe Berg who dropped in by parachute and was briefed by the Norwegian Resistance. It was confirmed that not only was the plant back up and running but the Germans had enhanced its heavy water production with an additional eight filtration cells. Just before dawn on 16 November 1943, 300 B-17 Flying Fortresses and B-24 Liberators of the US 8th Air Force headed for Norway. While half the force made diversionary attacks on Luftwaffe airfields at Stavanger and Oslo, 145 B-17s crossed the Hardanger Plateau, watched from the ground by Haukelid and Skinnarland. The aircraft dropped 700 bombs on the Vemork plant. A second wave of forty aircraft followed straight after. Only two bombs hit the target, however, and damage was limited but twenty-three civilians were killed. While not halting heavy water production in the way intended the raid convinced the Germans that Vemork was too exposed especially with the huge US bombing fleets now operating from British airfields. They decided to make plans for a heavy water plant in Germany and scaled back production at Vemork after first arranging for the transportation of all heavy water stocks to Germany.

On 20 February 1944, a train carrying cylinders containing all existing stocks of heavy water left Vemork for the coast where the cargo would be sent by ship to Germany. The train journey involved the trucks being decoupled and transported over Lake Tinnsjå on the ferry *Hydro*. Any attack on the ferry would risk reprisals against civilians but SOE ordered Haukelid and his team of saboteurs to press ahead with plans to sink the ferry with the trucks on board. Kjell Nielsen, working at the plant, was able to delay preparation of the consignment so that it could not leave until a Sunday when it was hoped that the ferry would have few passengers. The saboteurs, Alf Larsen, Knut Lier-Hansen, Rolf Sørlie and Haukelid, broke into the ferry terminal on the night of 19 February and planted explosives. At 10:30 the next morning the

Hydro, carrying fifty-three passengers, crew, guards and 14,000 litres of heavy water sank to the bottom of the lake, 430m deep. Eighteen passengers perished including a three-year-old child.[6]

Recent investigation of production records at Norsk Hydro and analysis of an intact barrel which was salvaged from Lake Tinnsjå in 2004 show that the containers held very low concentrations of heavy water which would have been virtually useless for nuclear research in Germany and seems to suggest that the plant had suffered much more damage from Operation Gunnerside than was thought at the time.

Chapter 11

THE SOVIET ATOMIC BOMB PROJECT

> 'Nobody has ever said, anywhere, that a nuclear bomb is not feasible, yet the opinion has been created that this goal belongs in the realm of fantasy.'
>
> Georgy Flerov, letter to Josef Stalin, April 1942

The story of the Soviet atomic bomb programme began with a little-known radium mine in Central Asia discovered in 1899 by V.A. Spechev at Tiuia-Muiun on the border of the Fergana Valley in present-day Kyrkyzstan. Spechev was disappointed not to find copper, which is what he was searching for, but the St. Petersburg Mining Institute analysed the tyuyamunite ore and found it to contain uranium and, more importantly, radium. A businessman and mining engineer, Khristofor Antunovich, began mining operations at Tiuia-Muiun in 1908 but after the Bolshevik revolution, private enterprise was supressed and the government imposed centralised planning and a command economy. Under a government initiative Vladimir Vernadsky and his pupils Alexander Fersman and Vitaly Khlopin set up a small factory on the site to extract the minerals.

Vernadsky would become a giant of Soviet science, having started the Commission for the Study of Natural Productive Forces (KEPS) in 1915 to support mineral exploration and prospecting. The Bolshevik government quickly began funding KEPS but unfortunately no significant sources were found and Tiuia-Muiun remained the only worthwhile radioactive-mineral deposit then known to exist in the whole of the Soviet Union and even then almost all of its output had been exported for much-needed foreign exchange. The marginal viability of the mine was shown by the fact that it produced only a single gram of radium for every 250 tons of uranium ore mined. The whole picture got worse when Germany, the main customer, went into catastrophic economic decline following the Treaty of Versailles and the Soviet government was forced to divert funding for scientific research into those areas that might be of more immediate relevance to the creation of a modern industrial economy. Between 1923 and 1936 a mere 20 grams of radium was produced at Tiuia-Muiun. Museum mineral specimens were tested for radioactivity and promising samples traced back to source. Searches for new mineral deposits started to bear fruit when another uranium source was discovered at Tobashar in 1927 but again it was low-grade ore and extraction of radium from the ore proved to be both difficult and expensive. The mine was forced to close after only two years of operations. Tiuia-Muiun followed suit in 1936 after flooding made further mining there impossible.

The Soviet Union, although a something of a political pariah, was still part of the worldwide scientific community which, on an individual level, was characterised by collaboration, open discussion and free exchange of knowledge. A number of leading Soviet physicists had done research abroad in the late 1920s and early 1930s, the most famous of whom was Pyotr Leonidovich Kapitsa who had spent time at the Cavendish Laboratory in Cambridge, at Marie Curie's Radium Institute in Paris and at Niels Bohr's Institute in Copenhagen. The Soviets were a little less enthusiastic than others and a bit paranoid about too much freedom of contact with the non-communist world and consequently tried to restrain Kapitsa by urged him to return permanently to work in the Soviet Union. Kapitsa did his best to ignore the political pressure but, during one of his frequent trips to Moscow, the decision was made for him when he had his travel pass revoked. It was a spiteful move since the government gave him little to do once he was confined. There was hardly any pure research going on with most scientists working in industry. He found 'no real scientific community' in Moscow where he felt that scientists were shown no respect and this made him 'bitter'.[1]

One of Kapitsa's contemporaries, the visionary Abram Ioffe, had been director of the Leningrad Physio-Technical Institute (Fiztech) since 1932. Over the next few years, he became the leading light in the development of Soviet physics, set up four research laboratories and organised annual conferences attended by scientists from all over the world including Frédéric Joliot and Paul Dirac. Scientists who would go on to play prominent roles in Soviet atomic research such as Igor Kurchatov, Yuli Khariton and Georgei Flerov all studied under Ioffe. At the time, communication and exchange of information about new discoveries was extensive and positively encouraged across the whole of the world's scientific community before the imminence of war forced institutions, under government directives, to clamp down on dissemination of research findings. Up until that point in the Soviet Union, as elsewhere, there had been much speculation that, with so much energy known to be locked up inside atomic nuclei, nuclear power could provide unlimited cheap energy for the world but there was also the recognition that it also carried the threat of it being used as a weapon of fearsome power. It was the military aspect of that research that now started closing doors to scientific exchanges. While that affected all countries, only the Soviets, who had enjoyed great success up until 1937, suffered from the sort of persecution that hit the LFTI when half of its scientists were swept up in the great Stalinist purges. Most were subsequently released but progress was stalled as a result and was slow to regain momentum afterwards, leaving many of them low on confidence and enthusiasm.

The Radium Institute which had been established in January 1922 by Vladimir Vernadsky, Vitaly Khlopin and Alexander Fersman had set out with the purpose of directing 'its work towards the mastering of atomic energy'. Already in 1922 Vernadsky had recognised the 'great revolution in the life of humankind' brought on by atomic energy but questioned whether it would bring 'good . . . or self-destruction'. He was clear that scientists should 'feel responsibility for all the consequences of their discoveries'.[2] The Institute would possess one of only two cyclotrons in Europe doing research on fission in 1939.

The influence of Vernadsky on Soviet nuclear research during the inter-war years cannot be overstated. Born into a politically liberal St. Petersburg family of academicians, he believed that science was a 'force for civilisation and democracy'. Never a strong supporter of the Bolsheviks, Vernadsky felt sure that by the late 1920s the whole communist movement would have lost its grip on the minds of the Soviet people which may well have been a common view amongst the intelligentsia, but Stalin looked elsewhere to establish his power base. As a man of science but also as a man whose intellectual curiosity took him to the fringes of philosophical debate, Vernadsky became increasingly interested in the teachings of the Jesuit philosopher and scientist Pierre Tielhard de Chardin who studied the way in which science was having an increasingly powerful influence on the human condition. Vernadsky could see that communism was stifling independent scientific thought, but he was confident that science, not politics, would ultimately free people from drudgery and underscore the advance of democracy.

Igor Vaslyevich Kurchatov was head of the Leningrad Physiotechnical Institute (Fiztekh), the home of nuclear physics research in the Soviet Union, and became 'the most important individual in the success of the Soviet Atomic project'.[3] His early career had included working at the Ioffe Institute in Leningrad, where with other aspiring scientists he had slept eight to a room in freezing conditions and had to cover his head at night 'to stop the rats from gnawing at his ears'.[4] He turned to nuclear physics in 1932 and gained a reputation among his peers for being bossy, argumentative and profane but, at the same time, good-humoured and described by the wife of a friend as 'like a teddy bear'. He had taken charge of the cyclotron particle accelerator at the Radium Institute in September 1939 to open up Soviet research into neutrons and uranium chain reactions even to the point of prising 10 tons of precious copper wire out of the grasp of the armaments industry for his electromagnets. Western experiments that had been reported in scientific journals were repeated and the results confirmed. Inspired by publication of the work being undertaken in Germany by Hahn, Strassmann and Meitner in 1939, Kurchatov had written a memorandum describing the military potential of nuclear fission and brought it to the attention of the Kremlin but he got little response. In August of the following year, he sent another report to the Presidium of the Soviet Academy of Sciences urging them to give some priority to further work on the uranium but again, his pleas fell on deaf ears.

Georgei Flerov was one of the few scientists in Kurchatov's department who had been given scope to keep atomic research alive. He had been, in a favourite phrase of Kurchatov, 'tasked' with replicating Hahn and Strassmann's 1939 experiments using pitchblende and a radon-beryllium neutron source, working deep underground in a laboratory in the Moscow metro system at the Dinamo station on Leningradsky Prospekt.[5] He worked at it for about eight months and eventually reported back to Kurchatov that he had indeed confirmed spontaneous fission, showing that the Leningrad group was as capable as any other at this time.

Few scientists in the Soviet Union thought that nuclear power would have any practical application within the foreseeable future. A report presented to the All-Union Nuclear Conference of November 1939 concluded that all research paths were 'fairly fantastic' given the huge quantities of uranium required to achieve a chain reaction.[6] Vernadsky, however, had been disturbed by an article in the *New York Times* his son sent from the US entitled 'Vast Power Source in Atomic Energy Opened by Science'. He petitioned the Soviet Academy of

Sciences for 'financial means for broad organised and rapid work in this area of first-class significance'. Any idea he may have had about accelerating Soviet atomic research to keep up with what seemed to be evidence of developments in both Germany and the USA, however, was hamstrung by the perennial and seemingly insoluble problem of finding enough uranium ore to get started. It was, therefore, not with any great sense of urgency that the Soviet Commission on the Uranium Problem was established on 30 July 1940 headed by Khlopin with Vernadsky, who by this time considered himself too old to lead, and Ioffe acting as deputies. Its goal was to work out a 'plan of measures which it would be necessary to realise in connection with the possibility of using atomic energy'. Its general direction of laboratory research was to look at methods of uranium isotope separation and enrichment and to organise research into achieving control of a chain reaction. A full assessment of costs and materials was drawn up. This was less than encouraging for uranium research when it was noted that extracting uranium was almost ten times more expensive than most other metals. The scientists were confronted by an insoluble dilemma. To produce enough uranium for experimentation, they needed to find a practical application for the ore to justify the expense, but to find that application, the commission needed more uranium for research.

The Fifth Nuclear Conference was held on 20–26 November 1940 when Kurchatov dampened the mood by stressing the 'enormous difficulties' of acquiring material and building reactors. Khlopin was equally pessimistic, complaining that the scientists working on the project were ignoring the 'needs of the present day'. 'Possible and practical', he said, 'were not concepts of equal value.' The 'beautiful dream' of uranium energy was 'a distant prospect'.[7] It seemed unlikely that Ioffe would put forward a strong case for investment and even Kurchatov did not have sufficient clout to put his case directly to his political masters. On a more practical level, the cyclotrons at the Radium Institute and LFTI currently under construction were to be completed quickly and new, more powerful, ones were planned for 1941. Despite all previous disappointments, the programme of exploration for new uranium deposits was to be re-energised immediately and construction of a heavy water plant was begun. August saw the drawing up of a plan for achieving a chain reaction using natural uranium and a good moderator in an experimental pile. It was not the best approach but one that had been chosen because of its lower costs and one that avoided the thorny problem of isotope separation. If Kurchatov had been able to press on with his plans and the Germans had not invaded, the first chain reaction might well have taken place in the Soviet Union because the Soviet physicists were not so far behind either the French, British or Germans in their thinking about nuclear fission.

After the Germans launched Operation Barbarossa on 22 June 1941, their armies moved quickly east. Soviet scientific institutions were hurriedly emptied as equipment was pulled back beyond the range of German bombers and personnel were drafted into the military or assigned to urgent war work. The LFTI, the Radium Institute, the Institute for Physical Problems and the Physics Institute of the Academy of Sciences, from Leningrad and Moscow, were all evacuated to Kazan some 800km from Moscow in August 1941. Kurchatov was firmly told to concentrate on practical military developments that could be rapidly integrated into the bitter struggle for survival of the Soviet state. Work on uranium was temporarily abandoned. From the first day of the German invasion, he had sworn not to shave until the Nazis had been defeated which meant that quite soon his flourishing beard gave him the appearance of an Orthodox priest.

Still Flerov pestered the authorities like 'a man possessed'[8] for more funding for uranium research. He wrote to Kurchatov on 21 December 1941 outlining some preliminary theoretical work he had done even including a diagram of a 5m-long gun-type mechanism for an atomic bomb. Although Flerov was unaware of the Maud report, the contents of this letter show that he was thinking very much along the same lines as the British. Kurchatov was laid low with pneumonia at the time and did not reply. He seems to have fallen into something of a depression, having become disillusioned with the miserable living and working conditions at Kazan where faltering electricity supplies had made research almost impossible. While he had found it easy to ignore Flerov's letter, he could not ignore what followed.

By March 1942 Gorsky had enough information to prepare a report for Stalin in which he recommended setting up a group to make a full study of what atomic research was going on in the West. Only a month before that, Sergei V. Kaftanov, Plenipotentiary for the extremely powerful Scientific Affairs of the State Committee for Defence, had been handed an exercise book containing formulas for calculating uranium chain reactions that had been taken from a dead German mining engineer called Vandervelde in the Ukrainian town of Taganrog on the Azov Sea. The implication was that he had been investigating uranium ore deposits in the area which suggested that the German bomb project was very active. The committee were panicked into reopening of the Taboshar mine and ordered yet another large-scale search for uranium deposits which, like the others before it, proved fruitless.

By now, Beria had become much more interested in the idea of a bomb as shown by his diary entry for 28 March 1942 that included a note on the very high costs that were likely to be involved in its development. He noted that he had been 'watching it' for about a year and now wanted to be quite sure of the accuracy of the intelligence before taking it to Stalin. It was not a good time to be diverting resources away from traditional military production, however, given that the Soviet Union was in a life and death struggle with German forces at Stalingrad but Beria thought that he could 'look into it' when 'we are able to breathe easier'. Beria mentioned it to Stalin who began to take note. 'We need a strong organisation', he told his chief, 'with full participation of industry. But maybe it is too early [for the atomic bomb].'[9]

Scouring Western scientific publications to see how scientists in the West had responded to his recently-published work on spontaneous fission, Flerov had been shocked to find that all current issues were totally devoid of any references not only to his own article but to all nuclear fission and uranium research. The names of Fermi, Szilard, Teller, Anderson, Wheeler, Wigner and others had disappeared from print. It was obvious to Flerov that the work of all of these famous atomic scientists had now been shrouded in a blanket of secrecy and so must be of great significance. While Kurchatov mulled this over, Flerov, always a prolific letter-writer, was sufficiently anxious to write to Stalin himself in April. His letter, with its similarities to the Roosevelt letter, again showed that interest in the nuclear issue was following similar paths on both continents. 'Dear Joseph Vissarionovich', he began '… a uranium bomb would be a revolutionary change in the technology of warfare' and concluded with 'unless we can raise the level of activity to equal that of the foreign scientists and even to surpass them in some areas, we will make a big mistake and voluntarily give up a winning position'.[10] He outlined a series of measures to resolve issues that had stalled uranium research but doubted that a uranium bomb could be delivered in time for use during the war. Nevertheless, he teased Stalin with the observation that 'the state which first solved the uranium problem would be able to dictate its conditions to the whole world'.[11] Thanks to Beria, this did not come as much of a surprise to Stalin.

It may well have been as the result of Flerov's persistence that order No 2352ss went out on 28 September 1942 from the State Defence Committee to Ioffe's Academy of Sciences to begin work on uranium fission with a view to producing electricity and report back also on the potential for making an atomic bomb. Flerov was released from the army to join the programme in July. Scientists were recalled from

Kazan, Ufa and Alma-Ata to Moscow where Ioffe appointed Kurchatov as scientific director of the Uranium Institute.

On 26 November 1942 an urgent directive went out from Moscow to its embassies in the West to trawl for more sources of intelligence on the atomic bomb programme and Flerov suggested direct communications with individual laboratories. NKVD agents already in place were supported by the parallel espionage agency of the GRU. The Soviets had been running spies in the US ever since the Bolshevik Revolution in 1917 and had a long-established *rezidentura* base of operations infiltrating long-term agents into such as the OSS and the Office of War Information. All in all, it was now clear to Beria that he had to act on the atomic issue and so he set up a sub-committee of the State Defence Committee in February 1943 to investigate the possibility of the Soviet Union joining in the race to make an atomic bomb, giving the operation the code name Enormoz. Kvasnikov was given special responsibility for the New York *rezidentura* but it was Barkovsky who brought the first results. He had established contact with a source code named Eric, an 'Englishman' who had initially approached the CPGB with details of both British and US nuclear research. Eric, in fact, was not British but the 'highly intelligent, good-looking and well-connected' Engelbert (Bertie) Broda, a half-Jewish Austrian physical chemist who had arrived in Britain with his doctor wife Hildegarde in 1938 having suffered imprisonment in his home country for his political activism.[12] Broda was first identified as Eric in 2006 when his security file was made public although his activities had been discovered in KGB archives in 1999 along with those of Persian, another spy who operated in the US right from the start, Broda had been a person of interest to MI5 in Britain but, even though his mail was routinely intercepted, he was never arrested even though he was closely connected with a group of the Austrian Communist Party that had been established in London.

For a time the Soviet agent code named Persian or Pers (previously Fogel) was thought to be either the physicist Phillip Morrison or Oppenheimer but he turned out to be an obscure civil engineer, Russell McNutt, who worked for the Kellex Company, a Manhattan Project contractor. But after providing information about the uranium enrichment facilities at Oak Ridge, Tennessee, Pers ceased to be of much use to the NKVD when he stubbornly refused to relocate from New York to America's rural heartland, where the fissionable material for the bomb was being made. Despite repeated entreaties from Pers' Soviet patrons, McNutt stood firm, citing his ailing wife as the reason for resisting the move.

The young Broda had scaled Alpine peaks with Hans von Halban when they were undergraduates together before entering the small, international yet intimate world of physics between the wars where everyone knew or was aware of everyone else. After he arrived in Britain, he worked for a time as a research scientist at University College London but was interned as an enemy alien in October 1939 although released after a few weeks thanks to strong representations from influential people such as the former Austrian ambassador to Britain, Sir Georg Franckenstein, and Sir William Bragg, President of the Royal Society. Despite his background and against the advice of the security services, Broda was taken on at the Cavendish where he worked on the Tube Alloys project. In 1940, he joined Chadwick's team made up mostly of foreign scientists including von Halban and Kowarski where research was ongoing into the heavy water problem. It was around this time that the Brodas' marriage broke up and Hildegarde was given custody of their son Paul who had been born in 1936. The recent entry of the Soviet Union into the war on the side of the Allies after the Germans launched Operation Barbarossa impacted on British-Soviet relations and made it increasingly difficult for MI5 to resist pressure from scientific bodies for the recruitment of foreigners even those who, like Broda, held openly communist sympathies. The British MI5 officer and later Director General Roger Hollis bemoaned the fact that, although 'there [was] a very definite risk' that Broda would pass on information to the Soviet Union, it was 'most unlikely' that he would be able to get proof of his complicity.[13]

When Barkovsky met Broda for the first time, Broda insisted that he would only deal with someone who had at least a rudimentary grasp of nuclear physics, so Barkovsky had to take a crash course by reading *Applied Nuclear Physics* in what little spare time he had. Not prepared to restrict himself to details of his own work, Broda took to forging keys so that he could extract material from the safes of his colleagues. Soviet archives also revealed a report of 1945 which described Broda as having 'supplied an enormous quantity of most valuable, genuine documents [both] American and British . . . on the construction of uranium piles'.[14] Despite offers, Broda refused all payment for his espionage work.

Many of the British nuclear scientists moved to work in Canada but Broda elected to remain in Britain, possibly to stay close to his son Paul or maybe because he had started a relationship with the photographer Edith Tudor-Hart that turned out to be a little more than just professional. Tudor-Hart was born Edith Suschitzky in Vienna and had studied photography at the Bauhaus in Dessau in 1929 under Walter Peterhans. She supported anti-fascist and communist causes from very early in her life and worked for a time as a photographer for the Soviet news agency Tass, reporting on the shocking living conditions, unemployment, disease and hunger in Vienna, London and the Rhondda Valley in Wales. Immensely 'vivacious, amusing, curious and gifted', she moved to England but her political activism brought her to the attention of MI5 who described 'Edie the foreigner' as 'timid' and of slim build, pale complexion, blue-grey eyes and blonde bobbed hair.[15] For her part in communist demonstrations and promotion of industrial action, which was illegal for aliens, she was labelled as an extremist and expelled on 15 January 1931. All this and she was still only 20 years old. When the Dolfuss government came to power in Austria, Suschitzky was arrested and imprisoned for her membership of the Austrian Communist Party (KPDÖ).

After her release, she married an English doctor, Alexander Tudor-Hart, whom she had known since 1925 and who had left his wife and two children to be with her. She accompanied her husband back to England where she worked for several British magazines including *The Listener*, *Picture Post* and *Lilliput*, also earning a modest income as a children's portraitist. In 1934, with the Austrian Parliament dissolved, trade unions banned and Vienna coming under the grip of the Nazis, making it a dangerous place for Jews, her father Wilhelm committed suicide. By 1935, the Tudor-Hart marriage was over and she set up a photographic studio in Haverstock Hill from where she documented through her work the appalling conditions under which many Londoners lived. Her husband went to Spain to work as a surgeon during the Spanish Civil War. She was recruited to the NKVD in 1935 by Arnold Deutsch, who was doing post graduate work at London University as his cover, whom she had known from her days in Vienna.

During this time Suschitzky introduced a young British student, Kim Philby, whom she had also met in Vienna in 1933 through Philby's wife Litzi Friedman, to Deutsch, who then recruited him. For a time she acted as a courier between Deutsch and Broda with whom she had a brief affair. Her flat became a meeting place for other 'persons interested in communist matters' and was closely watched by MI5.[16] She was pulled in and questioned over the Arsenal spy plot in 1937 owing to her connection with one of the Leica cameras involved in the illegal photographing of secret military documents at Woolwich but nothing was proved and she was allowed to go free. A year later, Tudor-Hart acted as liaison between the CPGB and a new Austrian émigré outfit of which Broda was a member; a role in which she excelled being 'unobtrusive, self-effacing and discreet'.[17] She was now actively engaged in recruiting potential agents for the NKVD, under the code name Edith, and it was in this role that she first met and recruited Kim Philby and was used as a courier for Guy Burgess. After the Molotov-Ribbentrop Pact of 1939, clandestine Soviet activity in Britain came under ever greater scrutiny and it was during this time that Tudor-Hart became a vital link between Moscow and its British spy ring. It was an uneasy time for many communists in the West seeing the Soviet regime getting into bed with the fascists, but for Tudor-Hart, the discomfort was compounded by having her photographic studio in London destroyed by German bombs which had a severe detrimental impact on her life. In May 1942, an MI5 report described that 'from her dress and habits, she does not seem to be in affluent circumstances'.[18]

In late 1942, Broda started passing information to Tudor-Hart about British and US nuclear research but soon he was handed on to agent Glan whose real identity remains a mystery to this day. The two met every two to three weeks over a period of months during which Broda betrayed 'thousands of pages of documents on British atomic research'. Because of his access to reports from the US, by August 1943, Broda was considered to be '[Moscow's] main source of information on [the Manhattan Project]'. Special privileges were afforded Broda by his Soviet handlers such as allowing him to call a meeting by marking a certain page of the telephone directory in a particular phone box.

Eventually forced to abandon photography, Tudor-Hart resettled in Brighton, opening a tiny antique shop there on Bond Street and living right above the shop. She died from stomach cancer 20 years later in a pauper's hospice. Her only child, Tommy, was a severely mentally disturbed child with various unconfirmed diagnoses, ranging from schizophrenia to severe autism. He was incarcerated in an asylum. When she died, her ashes were scattered by an employee of the home with no-one else present. Once she was of no further use, she was effectively disowned by the Soviet regime from whom she had never asked for a single penny for her work. This stunningly beautiful and beguiling woman, whose life had been punctuated by a series of disastrous relationships, would end her life not too far removed from the conditions of others she had spent so much of her life bringing to the attention of the world, but her photographic work was posthumously recognised through books and exhibitions.

The decisions concerning Soviet atomic strategy were taken at the end of 1942 when the Soviet Union was in the gravest danger, with both Stalingrad and Leningrad under siege and the prospect of relief through the opening of a second front remote. Many scientists challenged them as not only pointless but effectively detrimental to the national effort for survival. To a very large extent the Soviet scientists were approaching the project without any clear idea of what progress had been made in the West. They had a good idea about the broad pattern of research that was taking place but did not really know which of the competing experimental pathways were showing the best results. There was little optimism that an atomic bomb could be built in time to be used in the war but, taking the long view, the post-war world would clearly be a nuclear world and Stalin wanted to be sure that he was not going to be disadvantaged politically or militarily.

The problems Kurchatov faced at the new Uranium Institute were many. He did not yet even have laboratories in Moscow to pursue the pure research that would inform the whole programme, the cyclotron had to be brought up to new specifications and uranium and graphite of the right purity had to be obtained through as yet undefined links with industry and the central government. He spent days studying the Maud and Tube Alloys reports that had found their way to the Kremlin thanks to Fuchs. Whilst they did not give very much practical advice they at least indicated which processes Britain and the US had abandoned and so would save much time and effort for the Soviet scientists. The extreme usefulness of such intelligence encouraged Kurchatov to instruct 'Intelligence bodies to find out what has been done in America' with priority given to Berkeley especially in relation to plutonium and cyclotrons.

On 11 February 1943, the Soviet State Defence Committee published Special Resolution 2872, signed by Stalin, to begin work on the utilisation of atomic energy with the goal of building an atomic bomb. Kurchatov was put in charge with Mikhail G. Pervukhin as his deputy for this project with Vyecheslav Molotov given overall political responsibility. Kurchatov brought in Flerov, Isaac K. Kikoin, Yakov B. Zeldovich and Abram I. Alikhanov who began work without delay. They chose to build a thermal neutron reactor and studied methods of isotope separation by thermal diffusion and gaseous diffusion known about from the open literature before the war. The first important task was to recall scientists who had been dragooned into the armed services and who arrived in Moscow with little more than the clothes on their backs. A permanent armed guard was placed around their laboratories. A request went out to the US in February 1943 to purchase stocks of uranium nitrate and uranium oxide. Groves was in a fix. He did not want to give anything to do with atomic research to the Soviets but neither did he want to alert them to any work that was being done in the US by denying the request and exposing the value he placed on the material. In the end he complied.

Three main directions of research were identified by Kurchatov who proved to be 'a quiet but through manager'.[19] Achieving a chain reaction in an experimental pile using natural uranium was the priority but developing methods of isotope separation and bomb construction were also on the agenda. Alikhanov was to develop a uranium pile with heavy water as moderator. Kurchatov explored the use of graphite while Artsimovich was to take charge of electromagnetic isotope separation, Isaac Kikoin the gaseous diffusion process and Anatoly Aleksandrov the thermal diffusion process. Plutonium was recognised as being of theoretical interest but as yet was not considered to be a practical avenue of research.

For many months, research on nuclear fission in the Soviet Union which, at best, had only long-term potential, had been curtailed as the country poured all resources into immediate strategies to survive the German onslaught. But now in February 1943, things were changing. Kurchatov had returned to Moscow to set up a new research facility Laboratory No. 2 of the USSR Academy of Sciences (Lipan) under the supervision of the 9th Chief Directorate of the NKVD. Into this laboratory was fed all the information amassed by Kvasnikov, head of the NKVD's 10th Section of the Fifth Department which had been monitoring Western nuclear research since 1939.

Requests had gone out under the auspices of the Lend-Lease Agreement for the US to supply uranium and may well have been sent but none reached Soviet nuclear laboratories. Exploration for uranium within the Soviet borders had been generally unsuccessful with only a few minor deposits found. Kurchatov had written to Beria to deplore the 'completely unsatisfactory' state of affairs with raw materials and a total 'lack of unified leadership'.[20] The authorities, he said, were simply not taking the issue seriously enough. He knew enough about the Manhattan Project to appreciate that the gap between the US effort and that of the Soviets was enormous.[21] As early as July 1943, Kurchatov had received enough information from the US to get the big picture but the detail was missing.

At the end of October 1943, when he had heard of Niels Bohr's escape from Copenhagen to Sweden, Kapitsa wrote to him, 'inviting Bohr and his whole family to settle in Russia, where he would be given all the help he wanted for carrying on his scientific work in association with the Russian physicists'. Bohr collected this letter from the Soviet Embassy in London in April 1944, after returning from the United States with a message from Roosevelt on the control of atomic energy. Bohr sent a 'warm but innocuous' reply to Kapitsa and informed the

British authorities of the exchange.[22] Kapitsa's letter must have been written with the knowledge of the Soviet authorities or else he could not have promised what he did in his invitation to Bohr.

In April 1944 Laboratory No. 2 received much of the equipment that had been evacuated to Kazan and had in its retinue Yuli Khariton who had worked at Cambridge in 1928. It had eight departments which corresponded roughly to those in the Manhattan Project;

Soviet Union	**Manhattan Project**
Atomic Reactors, headed by I.V. Kurchatov	Chicago Metallurgical Laboratory
Isotope separation by gaseous diffusion, headed by I.K. Kikoin	K-25 gaseous diffusion plant at Oak Ridge
Bomb construction headed by V.I. Merkin	Los Alamos
Heavy water headed by M.O. Kornfeld	Research done in Canada
Electromagnetic isotope separation headed by L.A. Artsimovich	Y-12 calutron plant at Oak Ridge
Neptunium and plutonium headed by B.V. Kurchatov	Hanford
Cyclotron headed by L.M. Nemenov	
Theory headed by I. Ya. Pomeranchuk.[23]	

The Soviet programme was now focussed not on building a bomb with which to defeat Germany but on making sure that it had all the necessary resources to make sure that it would not be left out of the elite club of nuclear powers that would be competing for dominance in the post-war world. Three types of reactor were considered: a natural uranium reactor with graphite moderator, like Fermi's pile; an enriched uranium reactor with a 'light water' moderator; and a natural uranium reactor with a 'heavy water' moderator. Kurchatov understood the advantages and requirements of the different designs and settled on a uranium-graphite pile with lattice configuration. His first priority was to locate a source of uranium oxide; about 60 tons to be precise. Next was setting up a cyclotron to produce small quantities of plutonium for experimental purposes. Again the three main methods of U-235 purification – gaseous diffusion, thermal diffusion and electromagnetic separation – were explored. Any progress the team made, however, was dwarfed by the reported developments in the US and it was clear that with the current rate of progress, the Soviet Union would be severely disadvantaged after the war by negotiating from a position of nuclear inferiority against the Western Allies.

Chapter 12

THE JAPANESE ATOMIC BOMB

The military in Japan had shown an interest in nuclear research as early as 1934 when the navy sponsored a study to look at the work done by Fermi and whether it might lead to the development of a new 'super weapon'. Results of the study were not encouraging but the navy did not drop the idea completely and, albeit with little sense of urgency, invited Asada Tsunesaburo, a professor of physics at Osaka Imperial University, to simply monitor research being conducted elsewhere. What did come out of it, however, was the beginning of a mood of cooperation between the military and university scientists. Previously the military had not generally trusted the civilian scientists who had lived and studied in the West and who might have been exposed to anti-Japanese propaganda.[1]

The Director of the Aviation Technology Research Institute of the Imperial Japanese Army, Lieutenant-General Yasuda Takeo had since his days at Tokyo University taken a keen interest in Western literature concerning research in the field of nuclear physics. When scientific periodicals suddenly went silent on the subject, Yasuda drew the obvious conclusion that the West was working on a bomb and had clamped down on security. In April 1940 he assigned one of his aides, an artillery specialist Lieutenant-Colonel Suzuki Tatsusaburo, to prepare a full report on the feasibility of producing an atomic 'uranium bomb'. Suzuki called on Sagane Ryoichi, another young physics graduate of international reputation whose expertise lay in cyclotrons which he had used as a student at the University of California, Berkeley and the Cavendish at Cambridge. Upon his return to Japan, he had started work on building the 220-ton cyclotron at the Riken Institute of Physical and Chemical Research (Rikagaku Kenkyujo).

Suzuki reported back four months later outlining the broad avenues of research to be followed including the mathematics and practical issues surrounding the physical engineering challenges. The report was not considered confidential and therefore was widely distributed among both the military and private industry. Of fundamental interest was the availability of uranium ores which Suzuki thought would be sufficient from within the borders of lands currently occupied by Japan, such as Korea with its rich mineral resources and copious supplies of slave labour. The report drew attention to the potential for not only a bomb of immense power but also atomic power as a source of energy to produce electricity and to power ships. There was no mention in the report of a timeframe within which either objective might be realised. The most difficult issue surrounding the exploitation of atomic power was recognised as the containment of the chain reaction to maintain a stable reactor without it running out of control and exploding. Unlike Western scientists, who went for heavy water and graphite as moderators, the Japanese chose cadmium.

During the early summer of 1940, Yasuda by chance met Yoshio Nishina and a colleague on a train in Tokyo. The three briefly discussed nuclear fission research and subsequently Yasuda approached Nishina to study recent developments in the field. Nishina, 'a chubby man with a handsome smile and a short temper', was the country's leading physicist. He had worked under Ernest Rutherford at Cambridge and with Niels Bohr in Copenhagen and counted Albert Einstein among his friends. He currently had over a hundred scientists working under him.[2] As a consequence of his time in the West, Nishina was much more relaxed and informal with his staff which was at variance with the normal authoritarian style of the day in Japan. He had written at least four papers on nuclear fission by the start of 1941 and had earlier discovered the isotope U-237 at about the same time as researchers in the US.

In 1937, Nishina had learned that Ernest Lawrence at Berkeley was constructing a giant 60in cyclotron and he sent Lawrence a letter requesting assistance in the construction of a duplicate at the Riken. Lawrence agreed and parts for the large cyclotron began to arrive at the Riken early in 1938. In summer 1940, Nishina sent his junior colleague Tameichi Yazaki to Berkeley to learn more about construction techniques. Yazaki returned to Japan with blueprints for the machine as well as important technical information, but it remained incomplete when war broke out between Japan and the US in 1941.

It was clear that nuclear weapons research in Japan at this time had low priority since it was some months before Yasuda took action on Suzuki's report. When he did, he called on Suzuki to make contact with Seibin Ohkôchi, director of the Riken, to look at the possibility of using nuclear fission in a weapon and Ohkôchi subsequently assigned the problem to Nishina. Yasuda realised that he was unlikely to get approval from the military for major investment given the still-fanciful notion that a bomb might be produced 'at some time in the future'. Japanese imperial ambitions required a focus on more immediate goals and programmes that would return swift results. Furthermore scientists, particularly those at the universities, showed very little interest in participating in the war effort.[3]

The first major issue that Nishina addressed was how to purify U-235 from uranium oxide up to around 10–20 per cent which, from his perusal of Western literature, he considered adequate to make a bomb. After that he would have to work out the size of a critical mass to avoid what he incorrectly believed was the risk of accumulating too high a quantity which might inadvertently 'blow up Tokyo'.[4] Research shows that Nishina at this time was in many respects 'ahead of anyone else in realising that he may not need pure U-235 [to achieve] a relatively small critical mass'.[5] His initial objective was to build a reactor in which a chain reaction was allowed to take place under controlled conditions and through observation of that calculate the conditions required to make a bomb, the essential feature of which is that fission would only be allowed to accelerate into an explosion at a time and place of someone's choosing.

By the middle of 1941, however, Japanese-US relations had deteriorated to the extent that the military was now rather more receptive to the idea of a 'city destroyer'. The Riken research station in Tokyo, set up in 1917, included research laboratories, storehouses and refineries and was given funding for more research but rather less than was required to launch a full-scale project. Then, just before 8am on Sunday 7 December 1941, 353 Japanese carrier aircraft carried out a surprise bombing raid on the US Pacific Fleet in anchor at Pearl Harbor in Honolulu. Four US battleships were sunk and four others damaged. Three destroyers and three cruisers were also sunk or damaged. As a result, Japan and the US were at war.

The Japanese navy now rekindled its interest in atomic power for propulsion given the problem of supplying fuel to its fleet which was now expected to operate over vast stretches of the China Sea and Pacific Ocean. While interest in an atomic bomb was not great, it was thought that research into the fundamental issues at stake in both could be mutually beneficial. Captain Yoji Ito of the Naval Technical Research Institute (NTRI) had been following developments abroad. In the autumn of 1941, he had started a feasibility study on the possible military applications of nuclear fission for the navy. Like Suzuki, he consulted Sagane at Tokyo Imperial University and concluded that the US was probably conducting research on atomic energy and that the Japanese military would be well advised to do the same.

Ito, an electronics expert, had just returned from Germany where he had been studying radar on behalf of the Japanese army but his work there had made him aware of uranium research being done at the KWI. He brought together a committee composed of all the major nuclear physicists and electronic experts in Japan that held its first meeting days after Pearl Harbor when it was attended by several high-ranking navy personnel. It is important to understand that Ito's committee had been set up under the auspices of the navy and Ito was put in no doubt by the navy representatives that any results or conclusions he came up with were not, except where absolutely necessary, to be shared with the army. In general, at the start of the war, scientific research, especially in a military context, had been less well coordinated in Japan than in Western countries and lacked unified, centralised control but after the military setback of Midway, atomic research was given new impetus. Even so the navy, which was primarily interested in atomic propulsion as a long-term project, was still reluctant to divert significant resources to it.

The meetings noted that several tons of uranium were stored at Riken but there was no answer to the question of whether or not it was enough for research purposes. What seemed clear to all, however, was the extremely unlikely prospect of even the US, with all its resources and the many eminent nuclear scientists who had fled there from Germany before the war, being able to make an atomic bomb within the foreseeable future. However, even as scientists left the meeting with the impression that the military had downgraded the programme, navy chief Kitagawa, head of the Navy Research Institute's Chemical Section, was planning to bring in a new face and make available to him significant financial resources. This man was the flamboyant Professor Bunsaku Arakatsu of the Kyoto Imperial University, a personal friend and former pupil of Albert Einstein in Berlin in 1926 and who had later worked at the Cavendish. He had built Japan's first particle accelerator. His work was heavily influenced by Hahn and his team at the KWI and his own research had been encapsulated in a paper published in the *Review of Physical Chemistry of Japan* on 6 October 1939. In it he had showed that uranium fission would produce just the right number of neutrons that made a chain reaction, and therefore a bomb, a theoretical possibility. Arakatsu would become the navy's recognised expert in the field of atomic research and received 3,000 yen to begin his programme.

By the end of 1942, much had changed. In May, the Japanese navy had come up against the US fleet in the Battle of the Coral Sea which was technically a victory for the Japanese but had halted their advance. Soon afterwards, the war in the Pacific had started to turn against Japan. It was clear to the Japanese leadership that the war was going to last rather longer than they had anticipated and the idea of developing an atomic bomb started to look as if it might be a factor in deciding the outcome. Responsibility for the programme was put under the strictest security and handed over to the army aeronautical department under General Takeo Yasuda who was given a substantial sum to progress the agenda. Promising students, one of whom was Kunihiko Kigoshi, were given deferment from military duty to join the research team, a rare concession. Kigoshi, together with Masa Takeuchi, was asked to look at ways of separating out U-235 but first they had to find a supply of uranium ore.

Nishina's research group held its first meeting on 15 January 1943, to look at ways of separating out U-235. Despite the noxious chemicals involved, the thermal diffusion method seemed to them to present the least number of difficulties. At the same time, the army was becoming more interested in the idea of a 'super bomb' to the point where the Prime Minister and Minister of War, Tojo Hideki, called in Yasuda and ordered him to accelerate the research programme. A report in June concluded that a bomb of tremendous power was possible but evidence suggests that Nishina, like Heisenberg, was fixated on an explosion from a reactor gone out of control, i.e. a reactor bomb, rather than on a true atomic bomb. Even though Tojo promised to increase funding there was no move to combine the resources of the army and the navy who remained in competition with each other and still showed no signs of cooperating.

In summer 1944, however, the army and navy changed tack and began to discuss collaboration on nuclear research efforts. The failure of both to make substantial headway in the development of an atomic bomb on their own resulted in formation of the Army-Navy Technology Enforcement Committee (Riku-Kaigun gijutsu unyo iinkai) in autumn 1944. In the following year, after nearly a year of negotiations, a tentative agreement was reached in which the Navy Ministry agreed to take a supporting role in the collaboration.

The most serious obstacle to progress was the availability of uranium ore. It was all well and good giving the uranium programme a new title the 'Ni-Go Project' (Nishina-Project) and boosting funding, which is what Tojo did, but they had to go cap in hand to Berlin to beg for two tons of uranium ore and even then were virtually ignored by the Germans, who had little to spare. Although uranium in various forms had been utilised in numerous ways, from decoration to film emulsion to scientific research, supplies had never been extensive. The most voluminous supplies, such as those in the Belgian Congo, had accumulated as a by-product of radium mining but by 1943 all known sources had been commandeered by Germany and the US. The only thing to do was to search for untapped sources nearer to home.

Luckily, Satoyasu Limori, one of the chief researchers at Riken, had spent years searching for radium and knew of several sites in the Far East where small-scale mining was taking place. One such was close to Tokyo at Ishikawa and others not too far away in Korea and Malaya but none offered the sort of quantities that a rich vein of pitchblende would provide and transport, especially by sea, was hazardous across contested waters. There seemed no way forward until Limori heard of a mine at Kikune near Seoul that had shown promise before it had been closed down. When he went to investigate he found it contained fergusonite, an ore with about 8 per cent uranium but the fergusonite itself was only about 5 per cent of the total yield which made the uranium content of the ore about 0.4 per cent. It was not promising but it was pretty much all they had and all they could get was shipped for processing by Riken.

Nishina's team had originally concluded that an atomic bomb was theoretically possible, but not technically feasible. By the end of 1943, however, there was a great deal of evidence that the US was developing an atomic bomb, not least from the US media where *The Coronet* magazine had published an article in May 1942 entitled 'Uranium 235: can it Win the War' and in the following year the *Long Beach Independent* ran with 'Uranium Bomb Could Change Course of War'.[6] Western scientists had been prevented from publishing anything pertaining to atomic research but the free press was something that even Groves had been unable to rein in.

In February 1944, Nishina reported to the army that the Clusius tube was nearly ready but that production of uranium hexafluoride was proving extremely difficult. Furthermore, his large cyclotron was still not running at full capacity due to the lack of high-frequency-generating vacuum tubes that it required. Continually beset such technical difficulties, the Ni-Go Project progressed at a frustratingly slow pace.

By spring 1944, the Nishina team had succeeded in producing just enough uranium hexafluoride to begin isotope separation experiments. However, production of this highly corrosive gas remained at a low level. By early summer, Kunihiko Kigoshi had manufactured only about 170 grams for basic research. When the army officers returned to the Riken again in November 1944, the update they received from Nishina turned out to be quite discouraging. The Clusius tube was losing uranium-hexafluoride gas due to corrosion of the copper tubing and separation of the U-235 isotope was proving beyond their means. By February 1945, however, Fumio Yanazaki, a member of Nishina's team, thought he had enough enriched uranium to test with a neutron bombardment to see if he could induce fission. The test proved to be hugely disappointing. There had been virtually no uranium enrichment at all. Everything came to a halt, however, on the night of 13 April 1945 when the laboratory was struck by US bombs and burned to the ground but the cyclotrons were salvaged still in a usable condition.

There was no hint of surrender from the Japanese leadership and indeed few in Japan would have expected it. The war would go on and everything would be done to avert defeat and prepare for the last great battle. The traditional hostility between the army and the navy was set aside with both now cooperating on the atomic bomb project. Suzuki was ordered to relocate his research laboratories to Kanzawa on Honshu's northern coast in preparation for further relocation to the Korean mainland if necessary.

At some point just prior to the end of the war, the Ni-Go scientists were still struggling with the most rudimentary stages of isotope separation and concluded that an atomic bomb was still years away. On 21 July 1945, they concluded that while an atomic bomb was theoretically possible, it could not be produced in time for the war. All nuclear weapon research programmes in Japan came to an end with the destruction of the buildings which housed the isotope separation experiments. The Japanese wartime nuclear weapons project had never amounted to much and had been dwarfed by even the limited German effort in regard to the number of researchers, the amount of laboratory resources and availability of materials such as uranium and heavy water.

On 7 August 1945, the day after the US had exploded an atomic bomb over Hiroshima, Nishina was questioned by General Kawabe who asked if he could build an atomic bomb within six months 'assuming we can hold out that long?' 'Under present conditions', Nishina replied, 'six years would not be long enough. In any case we have no uranium.' It is not quite true to say that Japan did not have any uranium. It had, by 1945, enough uranium ore but probably very little in a refined state.

Search parties went out across the whole of south-east Asia looking for uranium deposits but the chances of success were recognised to be very low indeed so another attempt was made to get uranium from

Europe. Ambassador Oshima in Berlin was asked to make 'immediate investigation' of the possibility to export Czech pitchblende direct from the mines to Japan by air. This was a desperate idea given the huge distance over mountains and the risk of hostile intervention and after a few test flights that came to grief it was abandoned. The next idea was to transport it by submarine. This was not such a wild idea. Trade between Germany and Japan using transporter submarines between Kiel and Penang had been ongoing for a while. Rubber, coconut oil and rare minerals went from east to west while machine tools and industrial components came back the other way. At first the Germans baulked at sending large quantities but it was Hitler himself who intervened and authorised the release of two tons but there are no official records of any uranium actually having been received in Penang.

As the result of extensive negotiations between Oshima and a special Nazi commission *Marinesonderdienst Ausland* in December 1944, Germany agreed to send a shipment of advanced military equipment to Japan. The rationale for such a move is unclear but details of the cargo gives a clue. On board the 294ft-long submarine *U-234* that left the shattered harbour of Kiel on 25 March 1945 was, amongst other things, a crated Messerschmitt Me 262 jet aircraft, a Henschel He 293 guided bomb and a half a ton of uranium oxide. If the European war was all but over then Hitler seemed anxious to do whatever he could to prolong it in the East. The passenger list included the Luftwaffe specialists Lieutenant-General Ulrich Kessler, Colonel Fritz Sandrat and Colonel Earl Neishling along with civilian German jet and rocket engineers and physicists. Japanese military personnel on board were Lieutenant-Commander Hideo Tomonaga and Lieutenant-Commander Henzo Shoji. The U-boat had been specially fitted out for the long passage through the English Channel then down the length of the Atlantic, around Cape Horn across the Indian Ocean to Japan.

Somewhere in the North Atlantic on 4 May, the *U-234*'s commander Lieutenant Johann-Henrich Fehler picked up a partially garbled short-wave radio signal in English. For several days there had been radio silence from its Nauen communications base near Berlin and now Fehler knew why. The Soviets had taken Berlin and Hitler was dead. Six days later they heard the order to surrender. *U-234* was to surface, fly a black or dark blue flag and make for the nearest Allied port. A signal from Britain ordered it to Gibraltar but Fehler chose instead to go west. The Canadians hoped that it was headed for Newfoundland, a notion that Fehler did little to dispel but his track was actually taking him further south. He had opted for the US where the crew, after an open democratic debate, had decided that they would be better received. The Japanese officers on board demanded that they be allowed to commit suicide which they did by taking large quantities of Luminol after which they were buried at sea. When it was clear where the U-boat was heading, the US sent the destroyer *Sutton* out to board it and take it to Portsmouth, New Hampshire. Local radio reported that Kessler, wearing a long black leather greatcoat, highly polished leather boots and an Iron Cross 'strutted down the gangplank . . . casually looked around, saluted . . . and swaggered to a waiting bus', then posed for newsreel cameramen.[7] Kessler and some other German officers were taken to Washington along with eighteen metal cylinders. It is not difficult to imagine the consternation felt at Manhattan Project headquarters when they heard that a submarine designated *U-234* had been captured with a cargo of uranium ore. Did this signify a previously unknown isotope that the Germans had produced as part of an ongoing atomic bomb programme?

Much of the evidence of Japanese atomic research and development during the Second World War was destroyed or remains hidden, which has given rise to speculation about just how advanced it became. There were even reports of an atomic explosion carried out by Japanese scientists. Rumours had persisted ever since the respected US journalist David Snell broke the story, entitled 'Japan Developed Atom Bomb; Russia Grabbed Scientists', on 2 October 1946 in the *Atlanta Constitution*. Snell was a reputable journalist attached to the 24th Criminal Investigation Department in Seoul and in that role he clearly had access to Japanese officers and scientists. His source, who was referred to under the pseudonym Tsetusuo Wakabayashi, was a Japanese counter-intelligence officer but who was probably more closely concerned with Japanese atomic scientists working at Hamhung in Korea. This source told Snell that 'the Japanese [atomic bomb] programme was far more extensive, far more capable and far more successful than many would believe'. He also alleged that, just before dawn on August 12, a Japanese atomic bomb known as '*genzai bakudan*' (greatest fighter), exploded off the coast near Hamhung.

In 2014, Dwight R. Rider conducted a forensic examination to try and identify Snell's source and he came up with Suzuki Tetsusabaro as the most likely candidate. This immediately casts Snell's story of a bomb into doubt since, on another occasion, Suzuki was quoted as saying 'We believed in 1945 that we could build a bomb but we had to work much harder . . . I was confident at the time we could have built a bomb if we had better equipment.' Which is a clear statement that Japan did not have an atomic bomb in 1945.

Chapter 13

THE ALSOS MISSIONS

'I have been expecting you.'

Otto Hahn to Colonel John Lansdale Jnr., 25 April 1945

Otto Pash would not give up his hounding of Oppenheimer despite the lack of evidence that he was a security risk. Groves, who with Lansdale was perfectly happy with Oppenheimer, was growing tired of Pash and his constant, stubborn pursuit of the man. With German military defeats in Eastern Europe and expectations high that the 'Second Front' would be opened up some time in 1944, Groves saw an opportunity to rid himself of the troublesome Pash. He would assign him to a top secret mission to follow the invasion forces into Germany and seek out its atomic research centres. Authorisation of the mission had come from Roosevelt who had been informed by Bush that the US had 'no indication of the status of the enemy [nuclear] programme and [had] taken no definite steps toward finding out'. It was the US army Chief of Staff George C. Marshall who approached Groves on Roosevelt's behalf because it was generally felt that 'the existing agencies were not well co-ordinated' and that 'there was considerably more friction between the various intelligence agencies than . . . previously suspected'.[1]

Groves reviewed all that the US knew about the German programme and it turned out to be not very much. Much of what little they did know had come from an economics professor and financial consultant Dr Erwin Respondek who had close personal ties with the German industrial giant I.G. Farben, members of the Nazi Party and scientists at the KWI which gave him access to a wide range of political opinion and a number of well-informed sources who were involved with the German atomic programme. Respondek was one of a very few valuable intelligence sources for the US through his links to August Ochsenbein, a Swiss diplomat stationed in Berlin, and Sam E. Woods, the one-time commercial attaché at the US embassy in Berlin, who moved to the American consulate in Zurich after the US and Germany declared war on each other.

With the Gestapo breathing down everyone's necks in Germany, Respondek was forced to employ subterfuge to get his intelligence to Woods. In May 1943, a courier dressed as a priest would cross the Swiss-German border at Lake Maggiore to Locarno, go into a phone booth and dial the US consulate in Zurich. Woods was given a message to say that 'his friend' had something for him and would then make the 125-mile trip to meet the 'priest' sitting on a bench at the end of a pier. He would produce a Belgian lace collar as identification and the intelligence exchange would take place. The message Woods received was that Hitler was ready to adopt 'measures resulting in destruction on an unprecedented scale' and five million Reichsmarks had been allocated to atomic bomb research.[2]

Respondek prepared a synopsis of all scientific work being undertaken in Germany and gave Woods the names of the principal scientists working on the atomic bomb. What Respondek knew and what he told Woods but which was not acknowledged by the US government at the time was that I.G. Farben and the main Manhattan Project contractors DuPont had established a business relationship, similar to the one DuPont had established with ICI, in which they agreed to 'exchange scientific findings' as early as 1926 and that arrangement would still be in place in early 1945.[3] Dupont had tried to supply I.G. Farben with 'military propellants and military explosives' when Hitler had come to power in 1933 in clear violation of the Treaty of Versailles but no deal was struck primarily because of US Senate hearings in 1934 at which DuPont executives were obliged to line up to deny the existence of any such arrangement. Nothing daunted, as part of an agreement under which 'the two firms would regularly exchange the results of experiments conducted in their laboratories so that . . . no secrets would exist between the United States and Germany'.[4] DuPont continued to supply information to Germany about non-military processes such as the manufacture of synthetic rubber. Respondek had been instrumental in drawing up the agreements. Hostilities between the two countries after 1941 did not seem to affect this pact. It must be said, however, that there is no known evidence that DuPont ever informed I.G. Farben about atomic research going on in the US.

While German atomic research seemed to have concentrated on propulsion rather than a bomb in 1942 and they were not being particularly secretive about it, Respondek had told Woods that the

German Armaments Ministry, financed by Germany's General Electric Company (AEG), had set up a separate highly classified atomic project under Speer's personal supervision at Bisingen in southern Germany where a super-high voltage cyclotron was planned. This new facility was established and run by the Swiss physicist Walter Dällenbach, whom Rosbaud called 'worse than the Nazis', but the cyclotron was never finished.[5]

British Intelligence was not much help to Groves. They had deliberately avoided trying to infiltrate any agents into the German atomic programme which would have been almost impossible anyway. Any such agent, fully briefed for his mission, would if caught provide the Germans with too much information about the British effort. They relied on intelligence coming through from sources such as Respondek and Rosbaud whose work allowed them to keep track of the movements of German scientists and their relations with the sort of industrial enterprises that would be involved in any large-scale U-235 production. Furthermore, the British were most judicious about exactly what intelligence they were prepared to share. It is indicative of the way that atomic research was seen as of little value by the German security agencies that Rosbaud was able to continue publishing reports by Hahn and Strassmann on plutonium in *Die Naturwissenschaften.* It was one of the ways he found to keep the Allies informed about the German programme. This gave a broad picture of how the German programme was developing and that, as far as could be seen, was not particularly fast.

Philip Morrison, a young physicist with the Metallurgical Laboratory at the University of Chicago, had been asked to look at German scientific and industrial literature to see if anything might be deduced from articles. He compiled a report from articles that had appeared in the 1942 and 1943 issues of *Die Naturwissenschaften,* two physics journals and an inorganic chemistry journal. One actually included a three-page appendix listing the location of German physics laboratories, their specialities and some of their personnel. Relying solely on public sources, his two-and-a-half-page analysis estimated the amount of uranium ore accessible to the Germans and examined the availability of possible moderators for a reactor such as beryllium, heavy water and graphite. One puzzling observation was that seemingly Heisenberg and von Weizsäcker had been released from full-time atomic research to do other work. Another snippet of intelligence that alluded to German disinterest in atomic research concerned a reported conversation, presumably provided by Rosbaud, between Lise Meitner and Otto Hahn in Sweden in 1943 during which

Hahn told Meitner that the German fission research was not going at all well.[6]

Groves believed that only way to arrive at any degree of certainty in regard to the German atomic bomb programme was for the US to move intelligence-gathering activities into the front line of the fighting. Lansdale concluded that only by moving into those areas where research was actually being carried out could they get reliable information not least because of the danger that German atomic research facilities, documents and other key intelligence sources would be destroyed during and after battle. Retreating enemy forces were liable to take away with them or destroy documents and equipment and there was the ever-present danger of looting and gratuitous destruction by marauding front-line troops.[7]

With so little intelligence to work on, Groves chose an engineer he had previously worked with, Major Robert Furman, to head up a new intelligence section specifically tasked with getting intelligence on the German bomb. Furman went straight to Oppenheimer to get background and advice on what to look for. Oppenheimer told him to keep an open mind and not assume that any German programme would necessarily be run on the same lines as that in the US. There was no way of knowing if they might have found completely different answers to the problems of uranium enrichment or plutonium production. What was almost certain, however, was that buildings involved would be substantial, at the limit of the range of Allied bombers and would require very large energy sources. A number of simple operations might be attempted. For instance, if it was possible to take small samples of water from rivers and lakes close to large industrial complexes and somehow get them back to the US then they could be tested for traces of radioactivity.

Luis Walter Alvarez would later become world-famous for his discovery that a meteorite had struck the earth some 65 million years ago and led to the extinction of the dinosaurs but in 1943 he was at the Met Lab in Chicago busy devising ways of detecting nuclear activity in Germany. First he came up with a brief outline of what a German programme might look like and which components of that plan would easiest to detect. With Oppenheimer's help he described what a plutonium plant might look like and how it might be different from a uranium plant. He worked out that xenon-133, a radioactive gas that was emitted by operating nuclear reactors, produced distinctive gamma and beta radiation and it would be possible to build a filtering device into an aircraft that would then overfly a suspect German installation and, using activated charcoal, trap any xenon atoms in the atmosphere which could then be identified in the laboratory.[8]

This was all well and good but it was no substitute for boots on the ground and face-to-face interviews with those German scientists who might be involved. In his report of 7 March 1944, Furman told Groves that, despite British confidence that the Germans were pursuing only laboratory experiments, 'it has been determined that the enemy has the necessary material and is accomplishing important research work' and that 'a great deal of work other than the normal research is evidently taking place'. Another report concluded that the Germans were trying to 'deceive us about the extent and progress of their [atomic] programme and so cause us to relax the pressure on our own'. In other words everything that was known or conjectured about the German atomic programme that indicated its backwardness could easily be no more than a smokescreen to hide their real intentions and successes.

Karl Cohen, a chemist working under Harold Urey at Columbia University, came up with an assessment that given the head start German scientists had in 1940 it was possible that by the spring of 1941, they could have begun construction and it was safe to assume that a full-scale thermal diffusion plant could be completed by the autumn of 1943 and possibly much sooner. That could give them an atomic weapon by the spring of 1945. While a covering letter from Urey gave his opinion that there was no more than a 10 per cent chance of the Germans actually doing that, Groves and Furman were not about to take that risk.

It was assumed that the most competent German scientists and engineers had been working on an atomic programme with the full capacity of German industry at their disposal. With the Germans in full retreat on the Eastern Front there was also the problem of getting to the laboratories and staff before the Soviets did, not least to secure any equipment and research documents as well as taking personnel into custody to prevent the Soviets whisking them away to work on any atomic projects they might have going. For the same reason that the British had refused to send their own scientists into Germany, Groves was not about to compromise the security of the Allied nuclear programme by sending any of his own scientists on the mission but still he needed personnel with a scientific background who would know a cyclotron from a gas cooker and who would know what sort of questions to ask of any captured scientists.

All that was going to take time and a great deal of effort but Groves was impatient to strike a blow. Integral to any German atomic programme, he thought, would be the KWI in Dahlem just north of Berlin and it was here that he authorised a bombing raid on 15 February 1944. The suburb had few anti-aircraft defences and the laboratories suffered heavy damage. Unbeknown to Groves, however, most of the scientists had already relocated far to the south to the Black Forest region.

The Alsos mission was established in late 1943 under Pash's command to search for personnel, records, material and sites to evaluate the German atomic bomb programme and prevent their capture by the Soviets. Its headquarters was established in London with an original team of six scientists with support staff. In support of this Ricard Tolman had arrived in London in November 1943 with a view to setting up a joint Anglo-US atomic intelligence committee. Previous experience should have warned him that getting the cooperation of British Intelligence agencies would be far from straightforward. They barely helped each other. Tolman became so exasperated that he often left Robert Furman to deal with them. The British argued that the Alsos mission was unnecessary and got little resistance from Furman to this view but whilst they may have achieved their aims by scuppering cooperation, it was Furman who was the real winner. Neither he nor Groves had wanted British involvement in the first place, having no high opinion of MI6 which Furman categorised as 'skating rather thin' and positively frowning on the decidedly un-military habits of the rambling, scruffy head of the British negotiations Eric Welsh. They had made the gesture purely in response to political pressures and had happily allowed the British to shoot themselves in the foot with their intransigence. From now on if the British wanted to know what Alsos was doing they would have to peer over Furman's shoulder.

Pash had not been an obvious choice to head up the mission. He had no understanding of the scientific aspects of the mission and at times 'exhibited these credentials unnecessarily at times which caused embarrassment to those in the mission and showed lack of good judgment', all of which in itself did not bother Groves, because he had not chosen Pash for his scientific knowledge but instead for his intelligence experience, aggressiveness in the field and loyalty. A scientific officer, Goudsmit, would be seconded to Alsos who would be responsible for carrying out the scientific investigations and making the necessary reports. He would tell Pash where he needed to go, Pash would get him there and then the scientist would conduct his investigations. As it turned out, the relationship between Goudsmit and Pash proved to be perfectly amicable and the two became close friends despite Pash's usual prejudice against 'idiosyncratic longhairs' whom he expected to have to 'lead by the hand'.[9]

The first Alsos mission went to Algiers at the end of 1943 and then to Naples where Allied troops were already driving north through Italy. In late 1943 a mission similar in scope to Alsos, Project Larson, was set up by the OSS whose purpose was to kidnap Italian scientists and bring them back to the US. Within Larson was a small operation called Project Azusa, which had the specific task of interviewing Italian physicists to find out what they knew of the German atomic bomb programme. Assigned to Azusa was Morris 'Moe' Berg (Remus) who had been a Major League baseball player and was an accomplished linguist, two things that did not often go together, with a reputation for being temperamental and somewhat independently minded.[10] Berg was fluent in German, Japanese, French, Italian, Spanish and Portuguese and had some knowledge of several other languages. He had earned a law degree from Columbia University in 1928 and eventually took a position with a Wall Street law firm while still playing baseball. After several disappointing seasons with the Red Sox and two years of coaching, Berg gave up baseball and began looking for a way to get involved with the war effort. He was initially sent to South America but the work there bored him and he applied for a transfer to the OSS.

It was he who had parachuted into Norway to reconnoitre the Vemork heavy water plant prior to the major US bombing raid and now he had been seconded by Groves to the Manhattan Project. For security reasons, Berg was not made privy to details of the bomb project at Los Alamos but he was expected to have some rudimentary knowledge of nuclear physics. During the winter months, in preparation for his mission, he read Max Born's *Experiment and Theory in Physics* and studied quantum theory and matrix mechanics. When Rome was declared an open city on 4 June 1944, Berg flew to Caserta and then onto Rome which was being occupied by US troops. In Rome he went to 50 via Parioli, the home of Edoardo Amaldi, an experimental physicist who had worked with Fermi. It was one of the rare occasions that Berg bothered to wear a uniform but he thought it necessary to impress Amaldi. When Berg told Amaldi he had been ordered to take him into custody Amaldi said he had already been contacted by an officer of the US military and had been told to stay put. Berg agreed for Amaldi to take him to meet the officer in question and it was then that Berg met Pash. It was a noisy encounter. Berg explained that he had orders to deliver Amaldi to Alsos. Pash in no uncertain terms told the decidedly un-military Berg slouching around in a military uniform that he was looking at Alsos. Undaunted, Berg went round to see Amaldi the next day with candy for his children and settled down to a friendly conversation over lunch with him about his links with German atomic scientists.

Through Amaldi, Berg learned of the whereabouts of Gian Carlo Wick whom he also interviewed in his laid-back, friendly way, hoping to get information about Heisenberg's whereabouts. Neither Amaldi nor Wick could provide direct information about the progress of the German atomic bomb programme, but Wick, who had kept in close correspondence with Heisenberg throughout the war, did give Berg valuable information about the locations and activities of some of the German atomic scientists. He showed Berg a letter he had received from Heisenberg dated January of that year, in which Heisenberg revealed that his laboratories had been moved to a woody region in the southern part of Germany.

It has been said that Sam Goudsmit had been considered suitable for Alsos missions 'principally because he was available' and 'knew so little about the Manhattan Project that his capture by the Germans would not compromise it'[11] but it was also the case that he had known Heisenberg since their student days. It was not an uncontroversial appointment, however. Goudsmit had previously been hauled in for questioning by G-2 about a letter he had written to English colleagues in which he had openly mentioned Tube Alloys. It was not without some difficulty that he had been able to convince his interrogators that it was just a way of 'comparing notes' with other scientists. The general conclusion was that Goudsmit was a bit 'dopey, naïve and not a little unworldly'.[12]

Goudsmit flew into London on 6 June 1944, D-Day, and met up with Pash, just returned from Rome, who brought him up to speed. Pash took a team of thirty-three scientists and support staff into France on 9 August 1944 and joined the leading units of the Twelfth Army as it advanced toward Paris. Having entered Paris behind three French tanks as the first US unit to enter the city since 1940, they went immediately to the Collège de France where Joliot had his laboratory. Deterred at first by sniper fire from residual German units, they eventually found their man, who had been active in the French Resistance during the German occupation, to be cooperative but he was not entirely trusted despite Pash finding evidence that Molotov cocktails had been manufactured in his laboratory. In fact, Pash refused to accept that by staying in Paris and continuing to work in his laboratory all through the occupation, Joliot had not been a collaborator. Pash's information was that Joliot had been given 300,000 French francs by the Nazis toward his research in the reduction of heat formation during the atom-splitting reaction. He will not have been well disposed towards Joliot either if he had known, as well he might, of his far left-wing politics. 'I am afraid for my life', Joliot told Pash 'I shall be grateful for your protection.'[13]

He gave Pash's team the names of several German scientists, Kurt Diebner, Walter Bothe, Abraham Essau, Wolfgang Gertner, Erich Bagge and Werner Maurer, who had either visited or temporarily been working at his laboratory and who might have been working on a bomb but Goudsmit was later be convinced that he knew little about any German atomic bomb programme.

Goudsmit had spent his time going over the latest intelligence reports concerning the German nuclear programme before flying into Cherbourg on 25 August from where he was to travel on to Paris, join up with Pash and help with interviewing Joliot and his staff but had found that travelling through the chaos of northern France weeks after the invasion was anything but easy. Once there, however, he got down to business, sifting through the piles of documents Pash had found before being rushed off to liberated Antwerp to check up on any uranium that might have remained at the *Union Minière* warehouses. Although Antwerp was under British control Alsos was given permission to secure the offices and records and pursue their investigations without interference. Pash found 68 tons of yellowcake that was taken and shipped out to the US. Documents showed that another eighty-one barrels had gone to Paris before the German occupation. Pash followed that up and found that they had later been shipped out and hidden near Toulouse. He commandeered those also and sent them to Marseilles from where they went to Boston and then Oak Ridge. The remaining 1,000 tons known to have been shipped to Belgium from Shinkolobwe in 1940 was believed to be somewhere in Germany.

Meanwhile aerial reconnaissance flights were sent over the Joachimsthal mine in Czechoslovakia to ascertain whether or not it was in full production. Analysis of images by mining engineers decided that it was producing only a few tons of ore per day. Three Douglas A-26 Invader aircraft swept over locations in Germany where it was thought nuclear research might be going on. Special filters sampled the air as they passed but when it was analysed there was no trace of radiation. It seemed as if the Allies were chasing ghosts until flights over the Hechingen-Bisingen brought back images of a series of small factory buildings of identical design that had been built in double-quick time. It was known that Heisenberg had recently moved to the area. New railway lines and power lines fed the site and a nearby slave-labour camp was established but further analysis showed that it was a false alarm and the installations were for processing oil shale for fuel. The work of Alsos in a battle zone was always dangerous. Russell Fisher had been a professor at North Western University and served as part of the Alsos Team in Europe. On 7 March 1945 he advanced under fire onto the Ludendorff Bridge, the famed 'Bridge at Remagen', in order to lower three bottles into the River Rhine below to obtain water samples to test for radioactivity.

The real prize, of course, was Heisenberg. Groves had identified a scientist and close friend of Heisenberg, Paul Scherrer, a professor at the Zurich Federal Technical College, and it would be through him that he would work. Scherrer (Flute) had been an informant for the US through Frederick Loofbourow, an executive of Standard Oil of New Jersey whose frequent trips to Switzerland during the war had provided cover for their liaisons. In the spring of 1944, however, Scherrer supplied intelligence of a different sort. He told Loofbourow that the chemistry section of KWI in Dahlem had been struck by bombs and severely damaged, causing many of its scientists to relocate under the supervision of Otto Hahn to Tailfingen in Württemberg. Von Weizsäcker, whom Scherrer called 'extremely pro-Nazi', had gone to Strasbourg while it was believed that Heisenberg, 'a Nazi propagandist', was still in place at the KWI physics laboratories.[14] Scherrer also heard from Gentner that laboratories had also been hit in Munich, Leipzig and Cologne and Heisenberg was preparing to relocate also to Bisingen in southern Germany where a massive new and powerful cyclotron was being installed.

A few months after relocating to Bavaria, Heisenberg accepted an invitation from Scherrer to go to Zurich and deliver a lecture at the Swiss Federal Institute of Technology (ETH). Heisenberg had accepted perhaps as a relief from the dramatically deteriorating situation in Germany. Scherrer had told US Intelligence he would be meeting Heisenberg and would try to find out from him what developments there had been on the German atomic bomb. Groves saw that there might be more advantage than mere intelligence. In 1942, a plan had been hatched to kidnap Heisenberg, again in Zurich, and had even got as far up the command chain as Bush before it had been quietly abandoned as too risky. This time, Groves assigned Berg to travel to Zurich, pose as a scientist and get permission to attend Heisenberg's lecture. There he would be joined by Goudsmit, now greying and a little overweight but still with his 'dopey smile'.[15] Goudsmit had been under surveillance by G-2 but Groves was happy enough for him to be assigned to missions in Europe. He was particularly suitable for this one since he knew Heisenberg personally. Berg was to go into the lecture armed with a pistol, listen carefully and try to deduce from Heisenberg's lecture whether or not anything he said indicated that German nuclear research was sufficiently advanced for them to make an atomic bomb. If Berg was convinced beyond reasonable doubt that Heisenberg was anywhere close to achieving such an objective his orders were to shoot him dead on the spot. It was a tall order for a non-scientist to understand all the technical language of a revolutionary subject delivered in German, not his first language.

It is not certain that Berg and Goudsmit fully understood the extent to which their mission would contravene nearly every international law on wartime neutrality. Switzerland was extremely important in terms of intelligence during wartime. The authorities there were relaxed about espionage personnel, but any operation carried out that embarrassed the government would severely compromise diplomatic relations and endanger the whole US intelligence apparatus there. What Berg had been assigned to do would endanger Swiss-US relations significantly if it went wrong.

If the lecture went off without incident, Scherrer, quite unaware of all the details of Berg's deadly mission, would then arrange a dinner to which Berg and Goudsmit would be invited. Goudsmit, however, was hastily reassigned to go to Strasbourg and meet up with Pash to go through a mass of documents that had been uncovered. He was able, however, to have a brief meeting with Berg in Paris to go over once again the sort of things that Berg should be listening out for when Heisenberg spoke.

Heisenberg delivered his lecture on 18 December. Berg found himself seated behind von Weizsäcker who seemed to accompany Heisenberg wherever he went. It turned out that the lecture, as far as Berg could make out, was on S-matrix theory which had no military relevance at all.[16] After the lecture Berg introduced himself to Scherrer. 'Doctor Suits sends his regards from Schenectady', he said. A few days later, Scherrer arranged for Berg to attend his dinner party with Heisenberg and afterwards the two men found themselves walking side-by-side through the darkened streets of Zurich. It was the perfect opportunity for Berg but he had not been able to convince himself that the criteria had been met for him to carry out a cold-blooded execution.

It was assumed that at this point, now that it had been confirmed that there was no German atomic bomb, the Alsos mission would be closed down and all personnel returned to the US. In fact the manpower and resource allocation for Alsos were actually increased in the first months of 1945 when the mission was given its own table of organisation and equipment so that it would no longer have to borrow equipment from US combat units. By February, Alsos had established two forward bases at Aachen and Strasbourg to supplement the main mission office in Paris. The reason for this was that the mission had now shifted its operational intelligence-gathering operations to countering any attempt by the Soviet Union to build its own atomic bomb by seizing German uranium ore, locating secret laboratories and interning prominent German atomic scientists to ensure that they did not fall into Soviet hands.

Throughout the war, US scientists who had travelled to the Soviet Union reported that they found their Soviet counterparts 'unduly curious in their questioning . . . concerning [the US] work on uranium fission'.[17] Now that the Germans had been eliminated as an atomic threat Groves, Lansdale and Pash concentrated their full counterintelligence efforts on the Soviet Union. The first person to appear on their radar was Joliot whom they knew to have very strong left-wing views and made no secret of his and his wife's communist sympathies. It was feared that now all atomic research carried out in France would be immediately shared with Moscow through Joliot. Because of this Britain was urged to cease any further exchange of atomic intelligence with France and it was deemed necessary to take control of any German atomic resources that might come within the French zone of occupation. Likewise any institutions that were likely to fall into Soviet hands were considered a threat.

The Allied advance into Germany was slow and was forced to halt when the Germans counter-attacked in the Battle of the Bulge at the end of 1944. The Alsos mission did not enter Germany until 24 February 1945. Galvanised by the knowledge that the Soviet army was closing rapidly from the east, threatening to overrun one German facility that had been identified as producing thorium and uranium and knowing that there was no way that the American troops could get there before the Soviets, Groves recommended to General Marshall that the plant be targeted. On 15 March, 612 B-17 Flying Fortresses of the US Eighth Air Force dropped 1,506 tons of high-explosive and 178 tons of incendiary bombs on the Auergesellschaft Works just to the north of Berlin. The plant and all of its contents were totally destroyed. Groves hoped to disguise his motives by simultaneously bombing the nearby German army headquarters at Zossen.

The document find in Strasbourg had proved to be most illuminating. Goudsmit was reading reports dated just four months earlier which still debated 'the fundamentals of reactor design'. Recent letters referred to 'research work', not large-scale construction. There was an urgent request for two 'slide rules'. Few documents were stamped 'secret'. The overall picture of German atomic research was of a 'small, poorly-funded, part-time research project'.[18] Goudsmit concluded that the German programme had never progressed much beyond the stage that

Britain and the US had been at in 1941. His certainty, however, was not shared by his masters back in the US who, not being German-speakers like Goudsmit, had to wade through the Strasbourg documents as they were translated and were inclined to reach the same conclusion but were worried that the Strasbourg documents had not told the whole story. Pash was now hot on the trail of the German scientists and felt sure he was close to finding them. The hospital in Strasbourg seemed to him the sort of place that might have laboratory facilities the atom scientists might have used when their own laboratories were damaged. His instincts were sound. He found Rudolf Fleischmann, an 'extreme Nazi', and five of his assistants posing as medical staff. Fleischmann bluntly refused to answer any questions, but his office contained documents that Goudsmit was quick to scour.

Fleischmann's paper trail led Goudsmit to Hechingen in the Black Forest where he knew a number of German nuclear scientists had gone. Part of the Strasbourg haul had been a whole pile of letters between von Weizsäcker and Heisenberg on nuclear fission. One letterhead had Heisenberg's address (Weiherstrasse 1405) and telephone number. The Hechingen area was scheduled to come under French control but Groves was almost as determined to keep German atomic secrets out of French hands as those of the Soviets. He ordered that 'there could be no question but that American troops must be the first to arrive at this vital installation' and so set up Operation Effective to ensure that any nuclear research documents or raw material that were there would be secured. Pash had already hatched a plan to send in men of the 13th Airborne Division to secure the area ahead of his interrogators, but it failed to get approval. Instead, he followed a ground assault across the Neckar river and into Haigerloch through biting late spring snowstorms. Just outside the town of Hechingen, Pash's men found a cave barred with a locked steel door.

When the bunker laboratory at the Dahlem had been bombed at the end of 1943, the German authorities had begun to consider moving the work of the uranium project to safer locations further south. It was Gerlach's idea to move to the Swabian Alps region of the Black Forest, just south of Stuttgart and Tübingen. All personnel who would not be needed for the latest series of reactor experiments in Berlin, including assistant director Max von Laue, were gone from the capital by the end of the year to set up in a vacant textile factory in Hechingen. British raids on Hamburg in August 1943 had also convinced Paul Harteck to transfer his experiments in isotope separation from the University of Hamburg to Freiburg. Hahn, whose institute had been involved in fission studies, had gone to Tailfingen near Hechingen to look for alternative space also on 15 February which was lucky for him because on that night his institute was burned to the ground by British incendiary bombs.

Despite the air raids, Diebner remained working in Berlin where a model reactor was built in the bunker laboratory, completed at the end of 1943, using 1½ tons of heavy water and uranium plates. The main laboratory consisted of a water-filled pit, rapid air and water pumps in case of an accident, remote-control apparatus for handling radioactive materials, a ventilation and heating system and heavy-water tanks with a purification system. Other rooms included a workshop and several smaller laboratories, where further research on uranium and heavy water was conducted. As protection from aerial assault and radiation from the uranium furnace, the facility was surrounded by 2m of iron-reinforced concrete. To supply Diebner with high purity uranium oxide, uranium sheets and cubes for the *Uranmaschine* experiments conducted by Diebner in Berlin, Günther Wirths had set up an industrial-scale facility at the Auergesellschaft plant in Oranienburg that was pounded to rubble by the US Eighth Air Force on 15 March.

During 1944 altogether five significant experiments were run in Berlin but they were continuously hampered by shortage of raw materials and Diebner was forced to abandon the city and go to Stadtilm while Harteck's group went to Celle. At the end of 1944 the KWI physics department in Hechingen ran the final uranium machine B-VII experiment in February 1945 in a rock cellar in Haigerloch and it came very close to a self-sustaining chain reaction.

In the town of Leopoldshall on 17 April the search of a Wirtschaftliche Forschungs Gesellschaft facility uncovered over a thousand tons of uranium ore stored in barrels that looked as if it been standing untouched for years. Transport was hastily organised to take the ore first to England via Hildesheim and then to the US. On 23 April 1945, it was Furman and Lansdale who had found the sealed cave in Haigerloch. Pash was quickly on the spot and forced open the steel doors by threatening to shoot the caretaker of the café next door. Inside the cave he found a concrete pit about 10ft in diameter in which was a heavy metal container. The caretaker told Pash all he knew which is that it had something to do with a uranium machine although he had no idea what that was. A British scientific intelligence team, escorted by Lansdale and Furman, was called in to help the Alsos scientists evaluate the scene in which metal barrels full of uranium and heavy water had been discovered. The few German stocks of metallic uranium were found to have been buried in a field close by. Their conclusion was that the reactor was not big enough to have contained a self-sustaining reaction. They could barely contain their delight and immediately began dismantling the 'first German nuclear reactor that had ever been found'.[19] Analysis of the uranium residues, however, would later show that the whole experiment had lacked sophistication and was conducted with uranium that had barely been purified at all, suggesting that it was more like Heisenberg's reactor bomb than a working nuclear pile.

Pash pressed on to Hechingen-Bisingen which had just been taken by French colonial troops. However, a French team led by Joliot had beaten them to it, dismantled the powerful cyclotron that Dällenbach had been building and spirited it away to Paris. In a converted woollen mill Pash's team found offices and makeshift laboratories occupied by a number of important scientists and technicians including von Weizsäcker, Wirtz and Bagge. Heisenberg's office was there also but it was empty. On his desk was a photograph of Heisenberg with four other men one of whom was Goudsmit.

Twenty kilometres beyond Hechingen, in a town called Tailfingen, Hahn and von Laue were tracked down and taken into custody but not without fraught conversations with the French occupiers who, fortunately, had no idea of the two men's importance. Still there was no sign of Heisenberg. Lansdale tells of how he had found Hahn sitting at a desk in his office with a suitcase beside him. 'I have been expecting you', he said.[20] Hahn quickly handed over all of his research notes and showed Lansdale where their stocks of heavy water and uranium were stored in a cellar. The circumstances indicate clearly that the Germans had done very little and had not even constructed a working pile. Other captives were less cooperative. Bagge refused to talk about any of his work while Germany was still at war.

Before dawn on 20 April, Heisenberg had left his office at Hechingen and set out on his bicycle for Gammertingen, hiding by day from the many Allied aircraft that passed overhead and travelling by night across country that he had known well since his youth. He was also at risk from SS troops scouring the countryside for deserters whom they summarily executed. Heisenberg recalls the roads full of bedraggled German-uniformed young boys, who had been drafted in a last-ditch effort to stem the tide of Allied armour, many of them 'lost, hungry and crying'.[21] There were emaciated foreign workers too, freed now from their labour but starving and lost. After an intensely dangerous and harrowing journey, he arrived at his house in Urfeld three days after leaving Hechingen.

On 2 May, two days after Hitler's suicide, Pash and his team reached Urfeld where they were directed to Heisenberg's cabin. He was waiting for them and invited them in. He seemed pleased at the prospect of meeting Goudsmit again. In the meantime, a few German officers entered the town which, although vacated by the Germans, had not yet been occupied by the French. They wanted to surrender to Pash which was just about the last thing he wanted them to do. Having managed to persuade them to return the following day, Pash hurriedly headed back to the French lines without Heisenberg. The next day, obviously embarrassed and very angry at having been hoodwinked by Pash, the Germans sent in a hundred or more SS troops to hunt down the Americans. Finding none, they executed a number of villagers. When the French finally took control, Pash returned and took Heisenberg into custody.

At the same time, Gerlach was picked up in his laboratory at the University of Munich and Diebner was captured close by, while Paul Harteck was run to ground in Hamburg. As well as personnel, an operation headed up by Lansdale in a salt mine near Stassfurt seized close to 1,100 tons of uranium ore which was just a little less than the Germans had removed from Belgium in 1940. 'This' said Groves, 'would seem to remove definitely any possibility of the Germans making any use of an atomic bomb in this war.'

The prospect of actually achieving very much more of value that had not been already uncovered by the Allies now seemed very low. This was clearly in evidence in a report (*Rechenschaftsbericht*) that had been discovered written by von Ardenne in September 1944 for the German Post Office that financed his work. While the whole thrust of the 'aggressively self-confident' report was aimed at making a plutonium bomb it contains precious little detail of crucial issues such as critical mass. Von Ardenne completely overlooked the fact that a gun-type of detonator would bring together the fissile material far too slowly and the means of delivery is left untouched. Having started out with much optimism the report petered out into vague assumptions but at least it would have been something to show to Hitler who probably could not have made head nor tail of it anyway.

Of course, the Soviet Union had its own specialist search teams operating in Austria and Germany, especially in Berlin, to locate and remove equipment, material and documents as well as personnel that might be useful to their own atomic bomb programme. The Soviet team was headed up by Beria's deputy Colonel General A.P. Zavenyagin and included Flerov. German physicists who were taken by the Soviets included von Ardenne, Wirths, Werner Czulius, Robert Döpel, Walter Herrmann, Heinz Pose, Ernst Rexer, Nikolaus Riehl and Karl Zimmer.

The ten German nuclear scientists, Erich Bagge, Kurt Diebner, Walther Gerlach, Otto Hahn, Paul Harteck, Werner Heisenberg, Horst

Korsching, Max von Laue, Carl Friedrich von Weizsäcker and Karl Wirtz, who had been captured by the British and US forces in 1945 were taken to Farm Hall in Godmanchester near Cambridge, England and held there for questioning for six months. None of the men were prisoners of war because they had not been members of the armed forces and none of them was accused of having committed a crime. They were essentially 'guests' being held at 'His Majesty's pleasure' and given a modicum of comfort.[22] It was clear from their conversations that they had no idea why they had been collected together at Farm Hall and had no inkling of the Manhattan Project across the Atlantic. Clearly it was their scientific work that interested the Allies but their greatest fear and the one that probably encouraged them more than anything else to cooperate was the fear of being sent back to Germany, possibly into the Russian zone.

Transcripts of secret recordings of their conversations along with evidence uncovered by the Alsos missions were the clearest indications of exactly how far the German atomic weapons programme had evolved during the war. The men conversed freely amongst themselves and in one-on-one conversations, quite oblivious to the fact that every word was recorded in what was to become Operation Epsilon. Both sides of the argument over the complicity of the scientists in a Nazi atomic bomb have found evidence to support their cause but it is important to consider the context of each statement before evaluating it.

Heisenberg, says Paul Rose in his book *Heisenberg and the Nazi Atomic Bomb Project*, had no reason to change his view, which was widespread at the start of the war, that a bomb would require a huge critical mass. His concentration on a reactor bomb using U-238 had come about because of the seemingly insurmountable problems with enriching U-235 and plutonium was pretty much out of the question given that it could only be made as an end product of uranium fission. None of the scientists on Heisenberg's team was able to come up with any answers to these problems. It was only after hearing that an atomic bomb had been successfully dropped from an aircraft and detonated over Hiroshima that Heisenberg was able to look at the issues from a different perspective. The news was greeted first with scepticism then with awe. Here was evidence at last that an atomic bomb could actually work. It is a bit like looking at a chess puzzle and trying to work out the sequence of moves from the many that seem available to win the game and then seeing the final position and thinking, 'oh yes, of course, if that is how it turns out I can see that you must do this and then this and then . . . '. His whole approach to the theory changed fundamentally from that which he had expounded in the first days of captivity.

The detonation proved that the Allies had found a way of separating U-235 or had been able to manufacture plutonium, both of which pathways Heisenberg had abandoned in 1942 as impractical or impossible with the resources available to him at the time. From wracking their brains over theoretical and experimental problems, the solution to which seemed to lay far off into the future, the men suddenly became aware of the enormity of what they might have done had they been successful. Whilst appalled at the actual explosion, Hahn at least expressed relief that it had not been he who had caused it although, of course, he had been the one to start it all. Von Weizsäcker tried to imply that he had not wanted to work on the bomb as a matter of principle in the first place. Heisenberg blamed the political and military pressures put upon them to comply and said that it had left them little choice, adding that he was glad not to have succeeded in making a bomb. Von Weizsäcker then exposed another side when he boasted that of course the German scientists could have done it if they had been given more support although, having seen what happened to the KWI, he doubted that the factories would have survived Allied bombardment. Wirtz tried to occupy the high ground by saying that it was actually all a German discovery but they had refused to follow it through to its appalling conclusion. Gerlack collapsed in tears. It was feared that both he and Hahn would take their own lives.

The next day, however, von Weizsäcker felt sufficiently recovered to claim that the German team had only ever wanted to build a reactor for essentially peaceful energy production and it was the Americans who had exploited it as a 'ghastly weapon of war'.[23] They collectively issued a statement saying that by the end of 1941, all work 'was concentrated on [peaceful applications] for the production of heat and thereby to drive machinery'. It also, somewhat predictably, praised Hahn for his early work claiming that the bulk of it had been done after Lise Meitner had left Berlin and she had therefore not been concerned with the discovery.

German military research during the Second World War had in no way been a failure. Indeed its successes were many including high-speed jet aircraft such as the Me 262, TV-joystick remote-controlled flying bombs, proximity fuses and high-speed, long-distance submarines. All these are in contrast to its failure to deliver the atomic bomb. The country's advances in production of synthetic rubber, oil, fertiliser and explosives were all hugely successful and its factory methods copied by others after the war. The KWI combined private and public patronage to facilitate scientific research and KWG researchers, in general, received generous support and enjoyed the finest equipment with excellent staff. There had been a close relationship between the military and the KWI ever since the First World War, so much so that the KWI became a training ground for scientists in the service of the military. When Hitler came to power, however, German science and the KWI in particular came under fire for its 'Jewishness' and pacifist tendencies. Many scientists lost their jobs because of the Nazi purge of racially and politically 'unacceptable' civil servants and many scientific institutions lost most of whatever independence they had previously enjoyed.

Chapter 14

'LITTLE BOY' AND 'FAT MAN'

> 'The scientists will be held responsible, both by the public and by their own consciences, for having faced the world with the existence of the new powers.'
>
> Arthur Holly Compton, director of the Metallurgical Laboratory at the University of Chicago, 4 June 1945

Of course, having a bomb and using it were two different things. The argument for finishing the Manhattan Project and proving that a bomb was possible was strong from a political perspective but the question of using it as an act of war involved considerations beyond the political or military. The US set up special committees to discuss the issues. An Interim Committee reported back on 1 June 1945 with the recommendation that 'the United States should try to retain superiority of nuclear weapons in case international relations deteriorated'. They also suggested that the threat of a bomb would be unlikely to force the Japanese to sue for peace and only a demonstration of its power would be effective. Potential targets were specified by a bomb target group headed by Groves. The bomb should be used as soon as possible, without warning and against a vital war production facility surrounded by densely populated workers' houses.

The test of the plutonium weapon, named Trinity by Oppenheimer, was rescheduled for 16 July at a barren site on the Alamogordo Bombing Range known as the Jomada del Muerto or Journey of Death, 210 miles south of Los Alamos. The device was a plutonium implosion device shaped like a large steel globe. It used conventional explosives around a central plutonium mass to quickly squeeze and consolidate the plutonium, increasing the pressure and density of the substance to its critical mass.

To detonate the device, the explosives would be triggered, releasing a shock wave that compressed the inner plutonium and led to its explosion. The device was to be enclosed in 'Jumbo', a massive cylindrical steel container which acted as a containment vessel, because of concerns that the test would not be a success. If the bomb failed to explode, 'Jumbo' would preserve the bomb's rare plutonium for future experimentation. In the end so confident were the scientists that the bomb would work that the $12 million 'Jumbo' was not used.

A base camp was constructed with ample security measures but spartan living conditions. Many miles of roads were paved to transport materials to the test site and masses of electrical wires and cables were used to bring in the power that would detonate the bomb during the eventual test. Disaster almost struck and panic definitely reigned amongst the observers when the device was being winched to the top of the detonation tower and came partially unhinged as it began to sway. It was eventually righted and made its way to the top of the tower without further incident. On 14 July, just two days before the scheduled date of the Trinity test, Edward Creutz led a dress-rehearsal practice test without any nuclear materials which was unsuccessful and the device failed to detonate. Hans Bethe quickly discovered the cause and it was rectified by introducing a new component.

Three observation bunkers had been set up about 10km from the detonation site where scientists hoped to measure the amount of energy released. On 15 July the plutonium bomb was assembled and hoisted atop the 100ft firing tower. As predicted, wind and rain began to batter the Trinity tower on the night of 15 July. Many were worried that the test would be forced to be postponed for several days. This would be a serious setback for Truman if it happened. In Germany, he had already delayed his meeting with Soviet and British leaders at the Potsdam Conference until he knew the results from the Trinity test. He was right to be cautious because there were so many theoretical variables that not all could be predicted. There was a certain amount of scepticism over whether the bomb would actually detonate or not and if it did just how violent would be the explosion. Fermi, not entirely in jest, offered bets on whether the detonation would ignite the atmosphere and destroy the world. There were several plutonium bombs in various stages of assembly, enough to be able to spare one for a test. The uranium bomb was never tested because there was only enough enriched uranium for a single bomb and the weapon design was simple enough that the gun-type design was considered almost infallible.

Soldiers were posted in several nearby towns in the event that they needed to be evacuated. Groves placed a call to New Mexico

Governor John J. Dempsey explaining that martial law might need to be implemented in the event of an emergency at the site. On 16 July, a thunderstorm delayed the test, which was initially scheduled for 04:00, for an hour and a half. At 05:29:45, the bomb detonated with approximately 18.6 kilotons of power in a blast wave that spread out across the desert and the New Mexico sky was suddenly brighter than many suns. An orange and yellow fireball stretched up and spread while a second column, narrower than the first, rose and flattened into what would become a familiar mushroom shape, symbolic of its awesome destructive power. On the day before the start of the Potsdam Conference, the device exploded over the New Mexico desert, vaporising the tower and turning asphalt around its base into green glass.

After Truman received word of the success of the Trinity test his need for the help of the Soviet Union in the war against Japan was greatly reduced. Up to that point it had been very much in US interests to use the Soviet Union as a threat against Japan by tying down the Japanese Kwantung Army in Manchuria, further weakening Japanese resistance, but now Stalin was poised to strengthen the Soviet position in the Far East by promising to occupy the Japanese island of Hokkaido. Truman had not wanted the Soviet Union to be in any way involved in the post-war administration of Japan and now he had proof that the US did not need Soviet help to win the war and so could keep them at arms' length. Restricting Soviet influence in the Far East was one of the primary ambitions of the US and the atomic bomb would play its part in achieving that goal. On the evening of 24 July, Truman approached Stalin without an interpreter and, as casually as he could, told him that the United States had a new weapon of unusual destructive force. Stalin gave no reaction but said only that he hoped the United States would make good use of it against the Japanese.

A directive written by Groves, approved by Truman and issued by Secretary of War Stimson and General Marshall ordered the Army Air Force's 509th Composite Group to attack Hiroshima, Kokura, Niigata or Nagasaki (in that order of preference) as soon after 3 August as weather permitted. No further authorisation was needed for subsequent atomic attacks. Targeting now depended on which city was not obscured by clouds on the day of the attack.

The uranium 'Little Boy' bomb, minus its nuclear components, arrived at the island of Tinian aboard the USS *Indianapolis* on 26 July with its nuclear components coming separately on five C-54 cargo planes. The *Indianapolis* would be sunk by a Japanese submarine just days afterwards on 29 July. Now that Churchill had been defeated in a General Election, the new British Prime Minister Clement Attlee together with Truman, but crucially without Stalin's involvement, on 26 July at the Potsdam Conference issued a warning to Japan to surrender or suffer 'prompt and utter destruction'. Stalin had wanted to delay any ultimatum to the Japanese until Soviet forces had entered the war in the Far East. Three days later the Japanese leadership rejected the ultimatum.

The Alsos mission had proved fairly conclusively that fears of a German atomic bomb had been unfounded. They had got nowhere near building one and this worried many of the scientists on the Manhattan Project because the main reason for their own bomb had now been removed. There was no need for a nuclear counter-threat and there was no longer a moral argument for finishing the project. They had a fair point but it was not an argument that impressed a government which had spent such a colossal amount of money to get this far and was on the point of becoming the sole nuclear power in a world that was nearing the end of global conflagration. Building a post-war world based on US values, they believed, in the face of Soviet domination of Eastern Europe would give them enormous status and the power to call a halt to further Soviet penetration and as the new US Secretary of State James Byrnes said it might make them more 'manageable'. Furthermore, while the Germans had surrendered on 8 May, the war in the Far East was still being fiercely contested, with Japan showing no signs of negotiating a peace and a growing threat that the Soviets would take advantage of Japanese embattlement to pursue its own agenda by extending its power in the region. It was becoming essential to force a Japanese surrender before Stalin could muscle his way onto the scene.

The problem for the politicians and those like Oppenheimer who continued to support the project was that the world was not going to take a bomb seriously until its effects could be demonstrated and it risked the US having its bluff called. That could mean backing down from confrontations deemed insufficiently crucial to deploy it or be forced to use in under circumstances that were not entirely under US control and the consequences of which could not be calculated. There was also the expectation that the Soviet Union would now boost its own nuclear programme with the prospect of them acquiring a bomb. There was absolutely no way that the US could allow the Soviet Union to become the sole nuclear power in a post-war world. At Interim Committee meetings, with the bomb now ready for deployment, Groves had taken the opportunity to bring up the question of 'certain troublesome scientists of doubtful discretion and uncertain loyalty' but it was felt that all must remain in place for the time being.[1] What happened after that was another matter.

'Little Boy', the untested uranium bomb, was dropped from the B-29 *Enola Gay* piloted by Tibbets at 08:15 Hiroshima time on the calm and sunny morning of 6 August 1945. The sequence of events that followed was as follows;

- As the bomb was released two electrical pull-out plugs attached to the aircraft came loose and started a timer. This was designed to ensure that the bomb did not detonate prematurely, destroying the aircraft.
- After 15 seconds a second internal timing system took over.
- After another 15 seconds a radar altimeter was activated.
- When the radar sensed the correct height the firing switch closed as 0.9kg of cordite ignited and launched the hollowed-out uranium projectile down the gun tube with a velocity of 300m per second where it collided with the solid uranium target.
- One millisecond later a chain reaction was started and the bomb exploded almost immediately.
- The bomb contained 64kg of 80 per cent pure uranium but less than a kilogram actually underwent fission (1.38 per cent).

Forty-four seconds after release at a height of a little over 1,900ft above the city of Hiroshima, 'Little Boy' exploded with a yield estimated to be around 15,000 tons of TNT. This figure refers to the total energy release but only about half is in the destructive force, about a third is converted into heat and light while the remaining 15 per cent is radiation. Within minutes 90 per cent of all the people within half a mile of the blast epicentre were dead. Nearly every structure within one mile was destroyed and almost every building within three miles was damaged. Less than 10 per cent of the buildings in the city survived without any damage and the blast wave shattered glass in suburbs 12 miles away. A firestorm eventually engulfed over four square miles of the city, killing anyone who had escaped the blast. Estimates put the death toll from the blast and its immediate effects at 70,000 but the full death toll may have been up to three times that since many later died from the short and long-term effects of radiation poisoning.

At 03:47 on 9 August, the B-29 *Bock's Car* piloted by Major Charles Sweeney lifted off from Tinian and headed toward Kokura Arsenal, a massive collection of war industries adjacent to the city of Kokura but by the time they had arrived, the city was covered in smoke and haze. The pilot then diverted to his secondary target of Nagasaki, an industrial centre and major port on the western coast of Kyushu. The bomb was released and at 11:58 Nagasaki time 'Fat Man' exploded over the city at a height of 1,650ft with a yield estimated at 21,000 tons of TNT. The bomb containing a sphere of metal plutonium surrounded by blocks of high explosives that were designed to produce a highly accurate and symmetrical implosion. This would compress the plutonium sphere to a critical density and set off a nuclear chain reaction. The result was the fission of about 1kg (16 per cent) of the 6.19kg of plutonium. Almost everything up to half a mile from the epicentre was completely destroyed. All living things within half a mile died almost instantly and almost all homes within a mile and a half were destroyed. Estimates put the death toll on a par with that from Hiroshima.

The Japanese Supreme Council, six top members of the government who effectively ruled Japan in 1945, met at 11:00 on 9 August, almost an hour before 'Fat Man' exploded, to consider surrender terms having declined to discuss the issue only the day before. The background to the meeting was that during the previous few weeks a large number of attacks had been made on Japanese cities. On 17 July, Oita, Hiratsuka, Numazu and Kuwana had been attacked, with Oita and Hiratsuka more than 50 per cent destroyed. Kuwana was more than 75 per cent destroyed and Numazu was hit even more severely, with something like 90 per cent of the city burned to the ground. On 20 July three cities were destroyed including Fukui which was more than 80 per cent destroyed. Two days later Ichinomiya was 75 per cent destroyed. On 2 August four more cities were bombed including Toyama which was 99.5 per cent destroyed. On 6 August, only one city, Hiroshima, was attacked. In the three weeks prior to Hiroshima, twenty-six cities had been attacked with a third of them completely destroyed. The question arises as to why the Council had decided to discuss surrender hours only hours after refusing to consider it and one hour before the second bomb had been deployed. Why had the bombing of Hiroshima made such a difference to a country with more than four million men still under arms and a population inured to destruction?

It had been clear for some time to the Japanese that they would eventually lose the war but they had hoped to strike a peace deal either through Soviet mediation or by the threat of large US casualties if they attempted to invade Kyushu where the defenders were well dug in. Hiroshima had not changed their thinking in that regard. That strategy, however, was undermined on 8 August. The Japanese

Foreign Minister Togo had urged the Soviet Union to mediate on 7 August but Stalin had responded by ordering Soviet forces to attack Japanese forces immediately on three fronts in Manchuria. The possibility of Soviet mediation had evaporated completely. It was this that had precipitated the Supreme Council meeting which was called hours before the Nagasaki bomb exploded. Whilst the Japanese might have put up stern resistance to a US assault on Kyushu it did not have the resources to simultaneously defend its northern coastline from a Soviet invasion. The second atomic bomb exploded while the Council was in the middle of discussions. Japan now accepted conditional surrender based on the Potsdam ultimatum but Truman rejected that and demanded unconditional surrender.

On 10 August, Groves reported the next plutonium bomb would be ready for use on the first suitable weather day after 18 August. The presidential orders had implied that he was to continue dropping the weapons as and when they became available unless another presidential order halted the process. Only two targets remained and so Groves asked for additional targets but Truman ordered an immediate halt to atomic attacks while surrender negotiations were ongoing. On the same day Japan offered to surrender to the Allies, the only condition being that the emperor be allowed to remain the nominal head of state.

On 12 August, the US announced that it would accept the Japanese surrender, making clear in its statement that the emperor could remain in a purely ceremonial capacity only, but the Japanese were locked in argument amongst themselves over whether to accept the American terms or fight on. On 13 August conventional air raids resumed over Japan and later that day, the emperor called a meeting of his Cabinet and instructed them to accept the Allied terms immediately. Major Kenji Hatanaka and Lieutenant Colonel Jiro Shiizaki led a group of junior officers to try and take control of the Imperial Palace and impose martial law, but they failed to gain the support of senior officials and the coup collapsed. Hatanaka and Shiizaki committed ritual suicide in the grounds of the Imperial Palace

The Soviets, however, continued their assault and drove into northern China, northern Korea and the Kurile Islands. Stalin declared that 'southern Sakhalin and the Kurile Islands will pass to the Soviet Union' and ensure that the Soviet Union retained an outlet to the ocean.[2] Furthermore, Stalin asked the US to be allowed to send Soviet forces into Hokkaido to accept the Japanese surrender there but Truman flatly rejected the idea. Nothing daunted, on 23 August Stalin then ordered his commanders to occupy the northern part of Hokkaido.

Truman made it clear that he would not countenance Soviet boots on mainland Japan and Stalin hastily rescinded the order claiming that there had been a 'misunderstanding' over the Potsdam terms. It was a humiliating reversal for Stalin brought about not least because of the new US dominance as the first nuclear power and one that spurred him on to drive the Soviet Union forward to become the second one.

Ironically it had been Stalin's desire to prolong the war with Japan long enough for him to take advantage of the Japanese defeat that had hastened its surrender and robbed him of that chance. Had he waited before invading Manchuria, the Japanese probably would have kept on fighting even after Nagasaki and that would have given him a better opportunity to get a foothold on the Japanese mainland. He miscalculated by mistakenly thinking that Hiroshima would precipitate a Japanese surrender, which forced his hand on 8 August and meant that his ambitions for Soviet expansion in the Far East were only partially realised. It was, in fact, his decision to attack in Manchuria, not the atomic bombs, that brought the Japanese to the peace table with empty hands. The US, however, chose to justify its huge investment in the Manhattan Project and its decision to drop the two atomic bombs on Japan by claiming that it was those very bombs that had forced the Japanese surrender and saved a million US servicemen's lives by avoiding an invasion.

Within days of the Nagasaki bomb, Attlee convened a secret *ad hoc* committee, named Gen 75, to map out Britain's atomic policy. Back in November 1944 Chadwick, Cockcroft, Oliphant and Peierls had drawn up a proposal for a British atomic energy research establishment. The Tube Alloys Committee had endorsed this recommendation in April 1945 and it was now submitted to the Gen 75 Committee which recommended the creation of the Atomic Energy Research Establishment in September. The committee decided to write to Truman suggesting a complete review of nuclear policy. Truman was slow to respond but a meeting was scheduled for 9 November. At the end of the meeting a joint communique was issued expressing a desire for 'full and effective co-operation in the field of atomic energy between the United States, the United Kingdom and Canada' although Chadwick had already sensed that the gap between Britain and America was widening and the 'cohesive forces' binding them during the war were 'rapidly dissolving'.[3]

Meanwhile debate raged over whether to base the British design on U-235 or plutonium. Tube Alloys had done a lot of research on electromagnetic uranium enrichment which would form the basis of a uranium programme but scientists who had worked in the US were very much in favour of plutonium, based on its greater efficiency as an explosive, but they had no experience of the nuclear reactors that would be needed to produce it and not nearly enough understanding of how to extract it. In the end it was the financial argument that persuaded Gen 75 to build a single plutonium reactor at Sellafield. This decision was almost immediately called into question, however, when it became apparent that the Hanford reactors had encountered serious problems and had been shut down. Groves advised Britain against going it alone which infuriated the Gen 75 Committee, who resented being 'talked down to' and prompted them to authorise the building of a gaseous diffusion plant at Capenhurst. There was agreement between the British Chiefs of Staff and the Cabinet Defence Committee to disregard as 'dangerous and misleading rubbish' any suggestion that the Soviet Union would not develop its own bomb and also strongly asserted that the US could not be allowed to have a monopoly on nuclear weapons.

Rapidly backtracking on the 9 November communique, Truman was soon to declare that it was 'inadvisable' to 'assist the U.K. in the construction of atomic energy plants', pleading the need for 'international control of atomic energy through the United Nations'.[4] What came next was even more humiliating for the British. Wartime agreements had given them a half share of all ore coming out of the Shokolobwe uranium mine and even though it was still some way from making use of it, Britain had started to stockpile it for future use. A survey of US atomic weapons had revealed that it had only a single bomb ready for use and required just about all the world's available stocks of uranium ore to bring its arsenal up to requirements. The Marshall Plan had been drawn up to help the world return to economic health after the war and involved vast sums of money being invested to that end. It was made very clear to the British government that in order to get its fair share of this pot of money it was going to have to not only give up all claims to future Shinkolobwe production but hand over two-thirds of its current stocks of ore. Almost broke and worn down by six years of war, with no bargaining chips left, Britain had no choice but to comply.

The US conducted two nuclear tests, Abel and Baker, in July 1946 at Bikini Atoll, a coral reef in the Pacific from which the indigenous population of some forty families had been removed. These were very public demonstrations to test the effect of a blast on large ships. Even the Soviet Union had been invited to send two observers, one of whom was Simon Alexandrov. The first bomb was dropped from a B-29 that had been a support aircraft on the Nagasaki mission. It exploded at a height of 158m with a force of 23,000 tons of TNT but fell off target, much to the disdain of Alexandrov who made it clear that he was not impressed by the low level of damage inflicted. Baker exploded with similar force some three weeks later but under water this time and a third test (Charlie) scheduled to take place was called off.

In May 1948, the US again performed atomic tests at Eniwetok Atoll to study core design in an effort to increase the efficiency of the explosion and so reduce the core size, allowing them to make more weapons from a given stockpile of fissile material. In the absence of outside observers this time, the first bomb on Engebi Island exploded with a force of 37,000 tons of TNT with an efficiency of 35 per cent. Two more tests were carried out with varying success, but the overall results meant that plutonium bombs could be built with only half the amount of plutonium used in 'Fat Man'.

Having a nuclear arsenal and having the capacity to employ it effectively were soon seen to be two quite different things. Military analysts assumed that any confrontation with the Soviet Union would be over Europe and any deployment of atomic weapons would be several thousand miles from New Mexico. Activation of nuclear-armed Strategic Air Command bombers would take up to five days by which time a lot of ground could be lost. Even if the bombers could penetrate Soviet air space, the height at which they would have to operate at night made the prospect of accurate delivery of the weapon open to doubt. There was no guarantee that radar guidance would put the bomb closer than within two miles of its target. Crews were not trained to fly and navigate at heights of more than 25,000ft and when extensive trials were carried out to locate targets in simulated conditions, none were successful, in what the commander of Strategic Air Operations against Japan Major General Curtiss LeMay referred to as 'the darkest night in American military aviation history'.[5]

By the end of the war, LeMay had ordered the firebombing of sixty-three Japanese cities resulting in the deaths of more than two and a half million people. In Operation Meetinghouse on 9 March 1945, 279 B-29s had firebombed Tokyo, killing as many as 100,000 people which, LeMay argued, called for a sense of perspective on the scale of Hiroshima and Nagasaki. Of course what he failed to say was that on 6 August it had taken a single aircraft to deliver destruction and it was this as much as anything that brought a whole new dimension to aerial warfare.

In March 1949, LeMay had under his control some 120 aircraft capable of delivering atomic weapons. His War Plan I-49 envisaged using the entire US stockpile of 133 atomic bombs in a single strike against the Soviet Union, destroying as many as seventy cities. There is no doubt that the US had the means to deliver a first strike against the Soviet Union that would have resulted in something like three million deaths. It was confidently predicted that with the rate of bomb production, almost three times that number of bombs would be available within two years.

Post-war atomic intelligence against the Soviet nuclear programme was not seen as an immediate priority. American scientific and intelligence leaders assumed they had perhaps as much as a decade before the Soviets could build a bomb. The result was a breakdown of intelligence in all three aspects of the intelligence cycle. Collection was done piecemeal, giving nothing like a comprehensive analysis of the status of the Soviet atomic programme. Estimates were based mainly on wild speculation about what they assumed the Soviet Union would and could do based solely on American and German experiences and not in any way based on actual information from the Soviet Union. Consequently, both military and civilian policymakers were given the impression that the Soviet atomic programme did not pose an immediate threat.

Fuchs and Hall reacted differently to the destruction of Hiroshima and Nagasaki. For Fuchs, it was as if he had never really thought very much about the sort of devastation it would wreak, bound up and blinkered as he was with the fascinating challenges his work had presented him with over the years. Hall, on the other hand, had been under no illusions about the consequences of his work. Together with Greenglass, it seemed right to Hall and Fuchs that the Soviets should have full access to US research both to avoid US monopoly of the atomic bomb and to ensure that there would be a balance of power in the post-war world which would prevent any further use of it. Hall made it clear that he did not have 'an uncritical view of the Soviet Union' but later admitted that he had perhaps been a little naïve by not seeing the bigger picture.[6]

Chapter 15

'JOE-1'

> 'We have evidence that within recent weeks an atomic explosion occurred in the USSR.'
>
> US President Harry S. Truman, 23 September 1949

Soviet knowledge of the Manhattan Project was summed up in a memo from Soviet State Security to Beria in February 1945:

> The energy of an atomic bomb weighing about 3 tons would be equivalent to 2000-10,000 tons of chemical explosives, with destructive effects not only from the shock wave, but also the high temperature and the radioactivity, resulting in annihilation of every living thing within a radius of 1 km.
>
> . . .
>
> The Americans have enough active material for one or two bombs not as powerful, but still strong enough to have practical significance as a new form of weapon.

It is significant that the memo includes a note on uranium mining.

> Sources of uranium ore in Czechoslovakia are in the Sudetenland . . . 20 km north of Karlsbad.
>
> Our intelligence indicates that the British intend to conclude an agreement with the Czechoslovak government in London to exploit these sources of uranium ore.[1]

This indicated to the Soviets that Britain was pursuing its own nuclear policy but a consequence of British interest in the Czech deposits, now in Soviet hands, was that when its requests to the Czechs were rebuffed it also became clear to them that the Soviets were staking a claim which was the first real indication they had that Moscow had joined in the race for an atomic bomb also. Just how enthusiastically, competently and quickly they would do that now that the war was almost over would shock the world. Stalin asked Beria to deliver him a bomb in two years but even Beria could not obtain the necessary nuclear raw material, construct the vast processing and manufacturing facilities, build the device and test it in such a short time.

The advance of the Red Army west towards the heart of Germany had taken it to Joachimsthal and in March 1945 the Czech president Edvard Beneš signed a secret agreement giving the Soviet government rights to mine uranium ore and transport all existing stocks to the Soviet Union. That actually sounded better than it was. An intelligence source had reported that machinery at the mine was in such bad shape it could not be quickly restored. Neither were there sufficient rail cars to transport the ore. There was no sign of investment going into the site and the mines and laboratories were idle.[2]

What hope the Soviets might have had about capturing German nuclear scientists was soon shown to be forlorn. While a substantial number of Soviet atomic scientists had been seconded from Laboratory No. 2 to follow the troops into Germany, research in Moscow ground almost to a standstill. The mission would prove to be only partially successful, given that most of the German scientists who had worked on nuclear research during the war had been whisked off to England for interrogation, but it was also clear from the evidence that the Germans as a whole had not progressed very far. Where the Soviets had some success, however, was in tracking down stocks of uranium ore in Germany, eventually taking away some 300 tons of the stuff.

One of the German scientists taken by the Soviets was Nikolaus Riehl, who had been born in St. Petersburg but was brought up in Germany. He had worked with Hahn and Meitner in Dahlem and had extensive experience of working with radioactive materials at the Auer Company before the war. K.G. Zimmer was the first employee of the Auer Company to be captured by the NKVD and he was forced to reveal the whereabouts of Riehl, who had evacuated to Rheinsberg in Brandenburg with members of the staff and with the remains of the equipment after the destruction of the Oranienburg factory. He had tried to move west in the hope of being taken by British or US troops. Flerov went to Rheinsberg in mid-May 1945 with their hostage Zimmer in tow and immediately arrested Riehl there. They took him to the NKVD headquarters in Berlin-Friedrichshagen and then with his family he was flown out to Moscow at the beginning of June 1945. At the grim location of Elektrostal near Noginsk, Riehl was held in virtual captivity to work on uranium refinement for ten years where he employed German prisoners of war. What was left of his factory at Oranienberg after the bombing and was 'not nailed down' accompanied him 'at full speed'.[3]

Soviet troops entered Berlin during the last days of April 1945 closely followed by a search team, including Flerov. They found the KWI stripped bare of equipment but full of detailed files that showed what little progress the Germans had actually made there towards building a bomb. They were slightly more impressed when they went to von Ardenne's laboratories at Berlin-Lichterfelde. On 10 May, von Ardenne wrote to Stalin personally offering his services and all his equipment to the Soviet Union, an offer that was enthusiastically accepted. Within two weeks, Soviet technicians were dismantling von Ardenne's laboratory for shipment, along with all his staff, to Sukhumi on the Black Sea coast where they continued to work on isotope separation.

The Soviet project up to 1945 had been ponderous at best. There had been no special industrial-scale investment and seemingly little appreciation of the political consequences of the US demonstrating its atomic prowess until Hiroshima brought home to Stalin just how much had changed. Doubts about whether or not the bomb would actually work were replaced by a grim acceptance of its reality. It was no longer just a theory. Kurchatov's plea for 'extraordinary measures' in the face of the 'slow development' of the Soviet programme had been ignored by Beria in May. While trying to fulfil the brief Stalin had given him, Beria's almost total ignorance of the science of the bomb and his paranoia had made him suspect just about everything that Soviet scientists told him.

When news of the Hiroshima blast got to Moscow, it dampened the mood of celebration over the defeat of Germany. It was clear that the bomb now posed a clear threat to the Soviet Union and would severely restrict its ambitions in Europe. Stalin understood that 'the balance [of power had] been destroyed' and that now the US would use its atomic monopoly to impose its own agenda on Europe and the world. When he asked his scientific advisers how long it would take to reach nuclear parity, he was told at least five years.[4]

Ten days after 'Fat Man' exploded over Nagasaki, under State Defence Committee order 9887 of 20 August 1945 Beria was appointed to head up the Soviet Special State Committee on Problem Number One with Kurchatov as scientific director. Beria's brief was to oversee 'the construction of atomic energy facilities and the development and production of an atomic bomb'. It was an appointment that said more about Soviet politics than it did about Soviet science, but Beria would prove to be effective in the role not least because of his 'singular ability to inspire both fear and enthusiasm'. Widely regarded as 'the personification of evil', Beria also 'possessed great energy and capacity for work [with] intelligence . . . will-power and purposefulness'.[5] Stalin had been stung by US success in building an atomic bomb and laid the blame for Soviet failure to keep up directly at the door of scientists whom, he claimed, had not made their case for resources forcefully enough. As a result, he now promised them everything they needed to build a Soviet bomb as quickly as possible. It would have been more in keeping with Soviet management to specify a date but there were so many unknowns that it only an open-ended demand could be taken seriously.

Many atomic scientists working in the West made clear their opinion that there should be no complacency about Soviet capabilities to build an atomic bomb. The Soviet Union, they said, certainly had the capability to build the industrial separation and manufacturing plants and competent scientists to make them work. They cited the Soviet's long tradition of scientific excellence, giving Ioffe, Vernadsky, Khariton, Kapitza and Kurchatov as current examples of their expertise. Politicians, however, having a more isolationist and less worldly outlook, were much happier making the sort of joke about the Soviets that if they wanted to bring an atomic bomb into the US in a suitcase they would first have to invent the suitcase. All the best German scientists, it was assumed, had been brought out by the Allies with only a few second-raters ending up in Moscow. Truman doubted that the 'asiatics' could match US prowess. Groves, fully aware of the monumental effort that had been required to build the US bomb, told Congress that it would take the Soviets 20 years to build a bomb. However, what he failed to take into consideration, and indeed could not have done so, was that he simply had no concept of the huge extent of the security lapses that had occurred during the war. Because of this much of the experimental work Oppenheimer's team had been forced to do to establish the foundations upon which to build the industrial plant and the choices that could only be made after extensive research had been handed to the Soviets on a plate by the likes of Fuchs and Hall which had saved them an enormous amount of work.

In the Soviet Union, there was an influx of competent industrial managers into the project who had grown up in the age of Soviet industrialisation and who had vast experience of operating the armaments factories during the war, but this was to be a challenge on a scale that none had faced before. While some such as Peter Kapitza had worked abroad at the Cavendish, others emerged from seminars run by scientists to explain the technicalities of the programme no wiser than when they went in. Boris Vannikov, the Minister for Munitions, bemoaned the fact that he was 'hearing words for the first time, they are not in my vocabulary'.[6] The task they faced was the same as the one that had faced Groves in 1942, to build a bridge between the laboratory and industry but industry on a vast scale. All the issues that had challenged Groves now loomed before men such as Iakov Terletsky, who was scientific adviser, and Vannikov, who was the technical overlord but with the exception of security which had been a nightmare for Lansdale and a simpler task for the authoritarian Soviet regime.

The Canadian espionage network run by Zabotin was given a wake-up call from the newly-formed 'Department S' autonomous intelligence service that had taken over responsibility for all atomic espionage and intelligence from the GRU and NKVD. He received orders at the end of August to make every effort to acquire technical drawings and calculations from the Montreal atomic research centre through Alan Nunn May. He had shown some reluctance to be re-activated but was persuaded to come up with tiny samples of enriched uranium that were sent to Moscow.

All this threatened to come to an end when Nunn May was recalled to London which entailed hurried arrangements for him to make contact with a new go-between in a meeting scheduled for 7 October outside the British Museum. At the rendezvous, Nunn May, whose code name according to Gouzenko was Alek, would carry a newspaper under his left arm and use the password 'Best regards to Mikel'.[7] Gouzenko's debriefings, however, had alerted the RCMP to Hall's espionage activity and he was subsequently put under surveillance in Canada and then in London by the Special Branch of Scotland Yard. Nunn May never kept the rendezvous with his new handler and neither did he make any of the fall-back meetings either. It is believed that he had been tipped off by the British chief of counter-intelligence Kim Philby, who was also a Soviet spy. When the news of Gouzenko's defection was broken by the US journalist Drew Pearson who wrote of a 'gigantic Russian espionage network inside the United States and Canada', Nunn May was arrested and charged with breaching Section One of the 1911 Official Secrets Act. At the Old Bailey, he pleaded guilty to what the prosecution described as a 'squalid case' but he argued that the Soviet Union had not been an enemy at the time he had passed intelligence to them. The judge referred to Nunn May's 'wickedness' and the 'dreadful thing' that he had done. On 1 May 1946, he was sentenced to ten years' hard labour but released in late 1952. Zabotin was recalled to the Soviet Union where he, his wife and son were sent to labour camp in Siberia.[8]

In early August, Lona Cohen went to Las Vegas in preparation for a meeting with Theodore Hall. Cohen, born Leontine Theresa Petka, had been born in Massachusetts to a Polish immigrant family and became a member of the Communist Party after moving to New York. She had married another communist, Morris Cohen, in 1941 and began working for Anatoli Yaktsov of the New York *rezidentura* by stealing details of US weapons developments. In her meeting with Hall at Albuquerque, Cohen received several pages of notes and diagrams which she hid at the bottom of a box of Kleenex tissues. Despite Hall's assertion that he did not think he was giving the Soviets anything he had not given them before, these notes proved to contain a complete diagram of the first atomic bomb.

The Cohens were exposed after the arrest of Fuchs in 1951 and fled to the Soviet Union. From here they travelled widely and were able to acquire New Zealand passports in Paris, claiming to be Canadian nationals, Helen and Peter Kroger. Using these passports the couple were able to re-enter Britain and settle in Ruislip where they set up an antiquarian bookshop as a cover for renewed espionage activities. In the basement of their house, at 45 Cranley Drive, they set up a high-speed radio transmitter for communicating with Moscow. In January 1961, the couple were arrested for espionage, convicted and received sentences of 20 and 25 years but in 1969, the British arranged a prisoner exchange and the Cohens were sent back to Moscow where they continued training colleagues for intelligence operations.

Greenglass and his wife Ruth went to New York in September on a short holiday break and were met there by Rosenberg, now under pressure from Feklisov to increase his output as part of Beria's new initiative. Greenglass, who worked in the explosives division at Los Alamos, wrote out as much as he could about the explosive lenses and implosion calculations for the 'Fat Man' bomb design. This, together with information that Fuchs had handed over to Gold, seemed to provide a reliable pattern of intelligence that the deeply suspicious Beria could have some confidence in. The Soviets were beginning to develop a fairly comprehensive appreciation of the US bomb design. The problem now was that a number of their main intelligence sources were being returned to Britain as the post-war US agenda of discontinuing dual research with Britain was kicking in. Fuchs, like Nunn May, was being sent back home and would be denied any further access to the work at Los Alamos.

On Beria's committee, things did not go well between Kapitza and Kurchatov which, given Beria's antipathy towards the independently-minded Kapitza, did not bode well for him. Both Kapitza and Kurchatov had solid scientific reputations, both showed skill at working within the Soviet system to express their organisational abilities, each was absolutely fearless in the face of greater authority and both had experience of running large laboratories. Kapitza forced the issue by insisting that that all experimental results pass across his desk before going to the committee. It was easy for his enemies to stigmatise Kapitza for having spent 13 years at Cambridge and as a result, a whispering campaign against him gained traction and came to a head when, having got pretty much all the detail of the plutonium bomb from Hall, Beria now had to decide whether to copy it or strike out on a new Soviet design.

Beria asked for reports from both sides of the argument. Kapitza strongly favoured a new approach that would be cheaper and quicker than the US model and was so incensed that he wrote to Stalin threatening to resign if Beria favoured Kurchatov's view that copying 'Fat Man' was the way to go. There was no doubt now that Kapitza posed a direct challenge to Beria's authority but he was not one to back down. In his letter to Stalin, he even likened Beria to a musical conductor who knew how to wield a baton but could not read the music. Stalin passed the problem over to Beria. Beria demurred. The scale of Kapitza's courage was clear when he refused Beria's request to meet in the Kremlin saying he was too busy and Mohammed (Beria) had to come to the mountain (Kapitsa's institute). In the end Kapitza grew tired of the constant confrontation with someone of Beria's grisly reputation and again asked to be allowed to resign. Beria asked for Stalin's permission to arrest Kapitza, which would have ended very badly for Kapitza had his request been granted, but Stalin replied 'Don't you touch him' but at the same time fired Kapitza on 21 December 1945.[9]

Just before this the NKVD had made one last effort to exploit Kapitza's relationship with Bohr. They hatched a plan to send the young theoretical physicist Iakov Terletsky to Copenhagen in November 1945 with a letter of introduction from Kapitza, posing as a naïve young physicist eager to meet the revered scientist. The similarity with Heisenberg's 1941 visit to Bohr is unmistakeable and profound. Kapitza complied and wrote the letter, hoping that his phrasing and terminology might alert Bohr to the subterfuge, but found the whole thing distasteful. Terletsky was briefed on the mission by Beria personally. He must have been overawed by meeting Beria's assistant, an 'egg shaped fat man [of] repulsive appearance' who made disgusting sexist jokes before ushering him into an enormous room where Beria stood behind his desk with 'his head . . . round on top and his facial features . . . severe, without any warmth or smile'.[10]

Terletsky, who knew little about uranium and therefore would not be able to divulge anything that might put Bohr on his guard, engaged Bohr in a discussion about nuclear reactors in an attempt to discover something of what he knew about the US bomb project. At the meeting Terletsky was unable to get very much out of Bohr who was not deceived by the attempted subterfuge. While the NKVD reported that they had learned nothing new from Terletsky's meeting with Bohr, others closer to Beria and Stalin, for obvious reasons, had tried to give a more upbeat impression but the important outcome was that Bohr had declined, once again, to visit the Soviet Union and little more was ever said about the visit. The only really positive thing to come out of it was that Bohr gave Terletsky the first copy of the Smyth report that the Soviets had not seen but they soon got more copies from a variety of sources. US Intelligence became aware of Terletsky's approach to Bohr when Moe Berg heard about it from Lise Meitner who made it quite clear that she had no interest in going to the Soviet Union either.

Unaware of just how much information the Soviets had already got through their espionage activities, Groves had estimated that because of being 'backward in science and technology and lacked uranium' they would take as much as 20 years to build a bomb.[11] Truman, given alternative views, put the US lead at seven to ten years but that was still a significant period. The Acheson-Lilienthal report of March 1946 expressed the view that security against the bigger bombs that surely would result from a worldwide arms race could be gained only through international agreements aimed at preventing secret research and surprise attacks. Based upon this report Bernard Baruch presented a plan to the UN Atomic Energy Commission on 14 June proposing the establishment of an international atomic development authority that would control all activities dangerous to world security and possess the power to license and inspect all other nuclear projects. Once such an authority was established, no more bombs should be built and existing bombs should be destroyed. The Soviet Union rejected this as a crude device to allow the US to maintain its monopoly on atomic weapons. They responded with a proposal of their own. While the Baruch Plan had proposed that the US reduce its atomic arsenal by carefully defined stages linked to international agreement on control, the Soviets maintained that the bomb must be banned and all US weapons destroyed before meaningful negotiations could take place.

The Soviets knew from Fuchs that the US had very few atomic bombs which meant that the race for parity would not be as hard as it might have seemed at first sight. While the US would, in the foreseeable future, have a lead in that they were in the production stage, if the Soviets made a huge effort they could acquire their own bomb before the US advantage became overwhelming. Stalin thought there was little danger of large-scale war in the immediate future given the war-weariness of nations but that would not last. There was a narrow window of opportunity and he was determined to make the most of it.

Western analysis had it that Soviet victory over Germany in the East had initially given them a sense of national security but Hiroshima had stripped that away in an instant. That was somewhat over-simplistic given that Stalin knew that the US might have the technology and the means but it did not pose much of a nuclear threat. The symbolism, however, was stark and no state could sit at the table with the US as an equal. Stalin's strategy in the immediate post-war world, therefore, was to act as if there was no US nuclear capability and refuse to be cowed or intimidated. The US faced a dilemma. Having developed an atomic bomb partly as a political tool to help shape the post-war world in its own image it was finding that, with the Soviet Union in control of half of Europe and threatening to extend its control even further west, the bomb was having much less psychological effect than anticipated especially with the Soviet Foreign Minister Molotov openly boasting that the Soviet Union was well down its own road to having one also.

In October 1944 Donovan had recommended to Roosevelt the setting up of a permanent centralised intelligence agency placed under the direct supervision of the president to focus on national and not just military intelligence. While Roosevelt had been an enthusiastic supporter of the suggestion, Truman would later argue that it might exacerbate already existing rivalries among the various intelligence agencies. The OSS was formally disbanded on 20 September 1945 but early the next year, Truman took the first steps towards creating the Central Intelligence Agency (CIA) to 'assure the most effective accomplishment of the intelligence mission related to national security'.[12]

Rear Admiral Sidney Souers, the first Director of Central Intelligence (DCI), began to formulate a plan for the development of the Central Intelligence Group (CIG)'s capabilities for intelligence on the Soviet Union, but this was easier said than done. The Soviets operated within a closed police state with an effective counterintelligence apparatus. From the start, the US effort was thwarted by highly placed Soviet spies such as Kim Philby who forewarned Moscow of US attempts to infiltrate the Soviet atomic agencies. It may well be, however, that the CIA was simply not focused on the threat of a Soviet bomb. Had they been, they might have taken more note of a German scientist, Hans Adolf Krebs, who had headed up the KWI biophysics laboratory before being taken to the Soviet Union in 1945. Maybe it was because he had simply walked into the US sector of Berlin in 1946 with information about nuclear research in the Soviet Union, but it just seemed too good to be true and his report of having been interviewed for a job at a uranium refinery in Elektrostal where reactor-grade uranium was being produced in large quantities seemed like Soviet disinformation to panic the West into a rash move. Whatever the thoughts of the Berlin CIA analysts, the information had no discernible effect on US estimates of how long it would take for the Soviets to build an atomic bomb. Krebs went on to emigrate to the US and receive the Nobel Prize for Physiology in Medicine in 1953.

US scientific intelligence at this time was severely restricted by a critical shortage of qualified scientists in government service. Indeed, the whole scientific establishment in the US lacked young, qualified recruits since the draft board had not given deferment to science students whose whole academic training was disrupted by military service. It was still the case that scientists and intellectuals in the US were suspected by the right-wing establishment of communist sympathies which discouraged many potential recruits to science especially during the McCarthy witch-hunts of the House Un-American Activities Committee when, according to the *New York Times* as many as 50,000 scientists had been barred from government employment. This caused Truman to intervene and say that it was 'highly unfortunate that we have not been able to maintain the proper conditions for best scientific work. This failure has grave implications for our national security and welfare.'[13]

There was a lack of consensus about the extent to which the CIA should target the Soviet atomic weapons programme that was a consequence of the US being an open and democratic country. While many within the military and the intelligence community advocated a strong and concerted policy, others, mostly scientists and diplomats, were vehemently opposed and argued for the internationalisation of atomic energy and atomic weapons to prevent a war in which they may be used again. The US Navy, still bristling from the catastrophic intelligence failure that led to Pearl Harbor, called for 'secret agents' who will try to 'ascertain the rate of progress of foreign nations [the Soviet Union in particular] in developing any new weapons'. It was suggested that 'young American scientists, perhaps of white Russian descent', who would be willing to live and work in the Soviet Union for many years, might work their way into the confidence of the Soviet scientific community. Possibly some young Polish, Latvian, Finnish, Lithuanian or German scientists who would hate the Russians enough could join the American effort.[14]

Arguably the task facing the Soviet Union to build a bomb was, in many ways, much tougher than the one the US had been faced with in 1941. Whilst exhausted by war, lacking modern industrial facilities and technical expertise it did, however, have the benefits of a command economy with centralised control in which resources could be marshalled with great concentration. The government did not have to negotiate with industry over contracts, profit margins and labour disputes and the question of worker health and safety did not come high up the agenda. There was also the brutal reality of slave labour in the gulags that could be called upon for the heavy, unskilled work of mining resources and fuel. Of course, the Soviets also had the benefit of having ready-made solutions to many of the technical problems of the scientific framework for the processes required to develop the plutonium explosive or the intricacies of bomb design. All that had been given to them by Fuchs, Nunn May and others. Beria dominated the whole operation with his usual air of menace and threat but, unusually for him, he was careful not to weaken the project through excessive discipline or punishment.

Even before the end of the war, Oppenheimer and Compton were advocating a free disclosure of all information about atomic weapons in a 'co-operative understanding' with the Soviets so that the US could strengthen its moral position. This was reinforced after Hiroshima and Nagasaki. Even Bush expressed a desire to see 'full dissemination of fundamental information . . . including that of atomic fission'. Einstein called for 'a new type of thinking . . . if mankind is to survive'.[15] On 1 January 1947, the McMahon Act created the US civilian Atomic Energy Commission which had responsibility for nuclear power and nuclear weapons development and control. Groves was entirely against the idea, believing that it would 'invite disaster' since the Soviet Union could not be trusted to keep to any agreement. He still had enough of an intelligence unit to pull together enough disparate lines of intelligence to know that the Soviets were making full use of their captive German scientists and going flat out to build their own bomb. Debate raged over who should have responsibility for what and all the time the Soviet bomb project was gaining traction. Committees were formed all across the intelligence community, each trying to get the upper hand but the efficiency of the whole operation was neatly summed up when a report of July 1948 candidly admitted that it was 'impossible to determine the exact status [or date of completion] of the Soviet bomb'.[16] The January 1949 report would repeat the July 1948 estimates and again in July 1949 it would report that the earlier estimates were substantiated with 1953 as the most probable date for the Soviet bomb.

The physicist Henry DeWolf Smyth had been commissioned by Groves to write an administrative history of the Manhattan Project. It was subtitled 'A General Account of the Development of Methods of Using Atomic Energy for Military Purposes' and was published only days after the Hiroshima bomb. It concentrated on the physics of the project, which were already widely known, omitting almost completely the chemical and metallurgical aspects. This report was translated and widely distributed in the Soviet Union. Whilst not deemed particularly sensitive, some information in the report proved to be of great benefit to German scientists now working in the Soviet Union. Wirths, in particular, was able to refine his uranium extraction methods according to information in the report and boost the output of uranium up to a ton every day.

Soviet slave labour was used extensively during exploration for new uranium deposits at sites such as Taboshary and the extraction of ores from sites as widely scattered as Ukraine, Estonia and the Caucasus as well as Siberia and the Xinjiang region of China. Many of the deposits were in remote areas without communications so roads and rail networks were built. The most abundant deposits, fortunately, were found in Saxony in Soviet-occupied Germany.

Kurchatov, whose staff had increased seven-fold in the 12 months since January 1946, set about perfecting the graphite purification plant in readiness for building a reactor, F-1 (Fizicheski-1), at Laboratory No. 2 similar to Fermi's P-1 in Chicago. The design of F-1 closely resembled the experimental reactor at Hanford which indicated that it was the fruit of espionage. The 50 tons of uranium required to feed F-1 came from the stockpile of Congolese ore the Soviets had liberated from Germany. When the reactor was ready for testing on Christmas Day 1946, Kurchatov invited Beria to witness it in operation. It was a great disappointment to Beria that the only tangible evidence that anything was happening was the clicking of a Geiger counter measuring neutron production and his naturally suspicious nature made him wonder if he was somehow being fooled by these unreliable and potentially subversive scientists especially when, for his own benefit, he was prevented from entering the reactor room. When F-1 was boosted to high levels of output, its control room was situated in a building over 1km away for safety. Stalin was so impressed that he held a reception at the Kremlin on 9 January to celebrate the achievement.

Fuchs had now been activated again. His London contact Kuczynski was long gone back to East Germany but he had enough contacts still in London for a meeting to be arranged with Alexandr Feklisov (Eugene) who had been the Rosenbergs' handler in New York. They met at the Nag's Head public house close to Wood Green underground station on 27 September 1947 and afterwards Fuchs handed over documents on the production of plutonium. Shortly before leaving Los Alamos Fuchs and John von Neumann had jointly filed a patent application for the invention of a new scheme for initiating the implosion mechanism and this was all included. Fuchs and Kuczynski continued to meet every two or three months. Fuchs was now working at Harwell which was running a plutonium pile for research into both peaceful and military use of atomic power and was one of three main divisions in the section for atomic energy under the British Ministry of Supply. He had travelled to the US to attend a conference and had, ironically, been asked by the British to find out as much as he could about the US atomic project now that Britain was being eased out of its 'partnership' arrangement. Naturally everything that he discovered, which was not insignificant, went straight to Moscow as well as London. When a number of arrests of communist sympathisers at Harwell were made, Moscow feared that Fuchs would be exposed and made plans to bring him to join the Soviet atomic project just in case but they were never activated.

It had originally been decided to base the design of the Soviet bomb on the US 'Fat Man' bomb but physicists in Moscow had come up with an improved design which was half the size and twice as powerful. This, however, threatened to be sidelined by a new development after Fuchs had handed over theoretical plans for a thermonuclear hydrogen bomb on 13 March 1948. This new type of weapon promised to be at least ten times more powerful than any atomic bomb currently in existence. Whilst development of this weapon would extend well beyond the time scale for a plutonium bomb, it meant that a parallel line of research and development would need to be started up.

An area that had been heavily industrialised for wartime armament production had become the building site for much of the Soviet uranium project. Sverdlovsk and Chelyabinsk had good power supplies and communications but were sufficiently remote to ensure tight security.

Under the supervision of von Ardenne, a gaseous diffusion plant was built at Neviansk (Sverdlovsk-44) in the Urals while an electromagnetic plant was built close by at Severnaia Tura (Sverdlovsk-45) in 1946. A weapons laboratory (Arzamas-16), along the lines of Los Alamos, was set up in an environment that could hardly have been more different from the New Mexico desert. Khariton, who had been appointed by Kurchatov to direct atomic research, development, design and weapons assembly had chosen the small town of Sarov which was in a clearing surrounded by dense forest some 400km east of Moscow. At the centre was a monastery with a cathedral and white bell tower but all around were signs of the slave labour camps of both political and criminal prisoners that had been employed making casings for Katyusha rockets during the war. Scientists and their families were brought to live in the spartan conditions isolated from the rest of the Soviet Union but the presence of a large criminal fraternity within the labour camps made life far from comfortable for them. The town itself suddenly disappeared from Soviet maps.

An industrial-scale plutonium reactor (Annushka) was planned at Chelyabnsk-40 in the Urals which was the Soviet equivalent of Hanford in the US. In contrast to Sarov, this was an area of great natural beauty with mountains and lakes but again marked by the presence of 70,000 slave labourers who had built T-34 tanks during the war and would now be in continuous view to the scientists. The site had good road communications and ample supplies of water for cooling the reactors. Scientists moved in to supervise assembly of the reactor in early 1948. To mark the occasion, Kurchatov made a speech in which he expressed his hope that the town might 'become a monument to peace'.[17] A full-scale separation plant (Installation B) was producing pure plutonium dioxide by early 1949 and a third element, Installation V prepared the plutonium for use in a bomb.

Probably the greatest challenge the Soviets faced was gaseous diffusion. Fortunately, the Oak Ridge K-25 gaseous diffusion plant had been described in a series of articles giving broad descriptions of the plant that were openly printed in the US journal *Chemical and Metallurgical Engineering*.These articles may have revealed no classified material but information contained within them on design, development and its operating problems proved to be invaluable for Soviet designers. The articles contained information about diffusion such as 'When the cascade is first started up it must be allowed to run undisturbed until the proper concentration gradient has been established at each stage' and 'it is important to design the cascade with as little volume as practicable'. It also contained important information about the handling of the highly corrosive uranium hexafluoride gas.[18] Both von Ardenne and Hertz worked on isotope separation and by 23 February 1947 all problems had been dealt with sufficiently to go ahead with construction of a full industrial production facility (D-1) at Sverdlovsk-44 but when the plant came to be commissioned it was clear that quality control had been abysmal and Beria was not a happy man.

German scientists were hurriedly called in to deal with the mess. Slightly smaller than the Oak Ridge plant, D-1 was still very impressive. Beria made a personal visit to find out what was going on. Build-up of uranium deposits in the compressor housing threatened to lead to the formation of a critical mass with untold consequences. Leaks of uranium hexafluoride gas meant that significant parts of the plant had to be redesigned. The gun-type mechanism of detonation was out of favour and work on the uranium bomb, RDS-2, was put on hold in favour of an implosion bomb which required a lot of research into conventional explosives that was done at KB-11 at Sarov. As with the Trinity test, synchronised timing of thirty-two explosive lenses proved to be an immense challenge. The work was done at KB-11 alongside testing of the neutron source that would initiate the chain reaction.

Plutonium was now the explosive of choice and Reactor-A became the first to produce it, working at vastly higher power levels that the F-1 test reactor. This brought its own problems, with cooling now becoming a major requirement. If the plant was for power, the heat would have been used to generate steam but here, it was simply an unwanted by-product that was captured by a water blanket and poured back into the lake it had come from. The reactor, that weighed some 1,600 tons, was programmed to produce 100 grams of plutonium from every ton of uranium feed. From here it was taken to Chemical Factory-B where the plutonium was separated out and cast into weapons-grade hemisphere-shaped ingots.

Pressure was immense for Kurchatov to bring Reactor-A into production. Loading of fuel began in June 1948 but was halted after two months because of accidents that caused overheating and melting of aluminium casings resulting in 'a serious mess'. Safety was taking second place with serious radiation leaks and even the threat of an explosion. A report was scathing of Kurchatov who had 'ignored all safety rules and precautions (especially when nobody is looking) and personally [gone] into areas where the radioactivity is significantly higher than the specified norm' on one occasion exposing himself to '150 times the accepted maximum dose [of radiation]'. Workers were routinely exposed to high doses of radiation, even being trained to extract radiated uranium blocks and disassemble the damaged part of the reactor by hand. Trolleys transporting irradiated uranium from Reactor-A to Factory-B for chemical extraction often got stuck and had to be manhandled.[19] Whatever the costs, very soon the reactor had produced a kilogram of plutonium and Kurchatov reported that there was 'now . . . the real possibility of an atomic bomb in 1949'.[20]

Arzamas-16 was buzzing by the middle of 1948. All aspects of bomb production were progressing and a PE-8 heavy bomber was modified to carry the atomic bomb. The first fuel slugs from Reactor-A were delivered to Factory-B on 22 December 1948 and by March of the following year there was enough for a bomb. It was now time to find a remote test site which, given the Soviet Union's vast expanse, was not too difficult. A team from the Institute for Chemical Physics that had monitored the US Bikini test firings was called in to choose one. They plumped for an uninhabited plain in north-east Kazakhstan, 170km west of Semipalatinsk, near the Irtysh River. The test firing of the first Soviet atomic bomb, RDS-1, was authorised on 18 August 1949 but Stalin refused to put his name to the Council of Ministers resolution fearing that the test might be a failure in which case Beria would bear the brunt of recriminations.

A usable weapon would require a radar or pressure trigger just as the Hiroshima bomb had used but for the RDS-1 test, the weapon would be detonated on top of a tower and triggered remotely as had been the case with the Trinity test. At one point the tower had come close to collapse as its foundations shifted. Building a new stronger tower stalled the test firing for some weeks. The firing was scheduled for 06:00 local time on a cool and cloudy 28 August. When observers started arriving at 05:30 they were told that the test would be delayed until 07:00 because of the weather but instruments started indicating a build-up of neutrons threatening pre-detonation so the firing was brought forward to 06:30. Detonation took place as planned to muted celebration with one observer asking 'how much misfortune and death will this problem bring to civilisation'.[21] The steel tower had vaporised but the crater produced in the sand was very much smaller than the one at Trinity. The energy released was initially estimated to be in excess of 10,000–15,000 tons of TNT. Kurchatov and Khariton were awarded the honour of 'Hero of Socialist Labour', given a prize of 1,000,000 roubles, a Zis-110 motor car and a newly-built private dacha.

In 1946, the US had begun setting up a monitoring programme by weather aircraft to sample the atmosphere for indications of any atomic explosions but the remoteness of Semipalatinsk made it difficult to check on Soviet activity. However, an aircraft flying over the north Pacific on 3 September picked up traces of radioactive debris. A flight over Japan two days later detected more. The samples were flown back to California for analysis and the results were sufficiently convincing for Truman to be informed on 23 September and tell the American people that he had 'evidence that within recent weeks an atomic explosion [had] occurred in the USSR'.[22]

In an unconvincing statement similar to the one Groves had issued after the Trinity test when he said that there had been an explosion in an ammunition dump, Soviet news agencies responded with:

> It is well known that the Soviet Union builds large scale works: hydroelectric plants, mines, canals and highways that sometimes require new technology and big explosions . . . To the extent that these explosions often take place in various parts of the country, it is possible that they have attracted attention outside the borders of the Soviet Union . . . There is no basis for anxiety among foreign countries. The Soviet government, despite its possession of nuclear weapons, will stand by its unconditional resolve not to use them.[23]

By far the clearest demonstration of the dysfunction within the CIA and the failure of the whole US intelligence community was a report of 20 September, 1949, 'Intelligence Memorandum No. 225, "Estimate of Status of Atomic Warfare in the USSR"' which predicted a first Soviet bomb in mid-1953. This was 23 days after the detonation of 'Joe-1'.[24]

On 3 October 1952, the United Kingdom became the third country to test nuclear weapons after the United States and the Soviet Union. The first British test, code-named 'Hurricane', was conducted at the Montebello Islands in Western Australia. On 13 February 1960, France conducted its first nuclear test, code-named 'Gerboise Bleue' (Blue Desert Rat). The day marked the beginning of a series of four atmospheric nuclear tests at the Reganne Oasis, in the Sahara Desert of Algeria.

Appendix 1

THE SCIENCE OF NUCLEAR FISSION

An atom is the smallest unit of ordinary matter. Atoms are extremely small, typically around a billionth (1/10000000000) of a metre across. Every atom is composed of a nucleus and one or more electrons bound to the nucleus. The nucleus is made of one or more protons and a number of neutrons. While neutrons have no charge, protons and electrons have opposite and equal electrical charge. Protons normally repel each other and the closer together they come the greater the force of repulsion they must overcome in order to bind. However, when these particles are forced together in the intense heat and gravitational fields at the centre of stars, they cross a threshold beyond which another force takes control of them and instead of being forced apart they become bound together by an extremely powerful short-range 'strong force' in a process of nuclear fusion. The energy that is used to overcome the repulsion then becomes locked up in the atom. Atomic nuclei of lighter elements up to iron are created in this way. Elements heavier than iron are produced in the much more extreme conditions such as those that exist when massive stars explode and create intense gravitational fields in supernovae. Other elements such as lead are formed largely through the radioactive decay of heavier elements.

Atoms are defined by their atomic number (number of protons) and structure which in the case of the simplest element, hydrogen (atomic number 1), consists of one proton and one electron. Examples of other elements are carbon (atomic number 6 and commonly has 6 protons, 6 neutrons and 6 electrons), oxygen (atomic number 8 and commonly has 8 protons, 8 neutrons and 8 electrons), iron (atomic number 26 and commonly has 26 protons, 30 neutrons and 26 electrons). Naturally occurring heavy elements such as uranium occur in a number of forms and are much less stable than lighter elements. Uranium (atomic number 92) has 92 protons and 92 electrons but may have anything between 141 to 146 neutrons. Each form of uranium with a particular number of neutrons is called an isotope. For instance, an atom containing 143 neutrons is the isotope called uranium-235 (92 protons and 143 neutrons). The most common isotopes are uranium-238 which accounts for over 99 per cent of uranium on Earth and uranium-235 which occurs naturally in low concentrations of a few parts per million in soil, rock and water and is commercially extracted from minerals such as uraninite.

In 1934, experiments led to the discovery that heavy atoms such as uranium were much less stable and could be broken apart into lighter elements by bombarding them with neutrons in a process called nuclear fission. This resulted in a release of some of the binding energy. When the uranium nucleus splits it forms two smaller elements. In the process of splitting, however, a great amount of thermal energy as well as gamma rays and a small number of neutrons are released. Furthermore, empirical evidence shows that after the uranium has split, the total weight of all the fission products is marginally less (ca. 0.1 per cent) than the weight of the original uranium atom. The matter that has been 'lost' has been converted into kinetic energy according to Einstein's famous equation $E = mc^2$. It should be noted that the concept of nuclear fission did not arise directly from Einstein's work and the great man played only a very minor and inconsequential role in the Manhattan Project.

When a neutron passes near to a heavy nucleus such as U-235, the neutron may be captured by the nucleus and this may or may not be followed by fission. If the neutron is captured it is added to the uranium nucleus to form a new compound nucleus. In certain cases, depending on the velocity of the passing neutron, the initial capture is rapidly followed by the fission of the new nucleus. Low-energy (slow or thermal) neutrons are able to cause fission only in those isotopes of uranium and plutonium whose nuclei contain odd numbers of neutrons. For nuclei containing an even number of neutrons, fission can only occur if the incident neutrons have high energy. Newly-created fission neutrons are in this category and move at about 7 per cent of the speed of light (21 million kilometres per second), while moderated neutrons move a lot slower, at about eight times the speed of sound (2,700 metres per second)

The probability that fission will occur is described by the neutron cross-section for that reaction which, in simple terms, is the area surrounding the target nucleus and within which the incoming neutron must pass if the reaction is to take place. The cross-section varies depending on the velocity of the passing neutron. It was the calculation of this cross-section in relation to neutron velocity that was to take up much of the theorists' time when developing atomic power. Most of the energy released is in the form of kinetic energy which is almost immediately converted into heat.

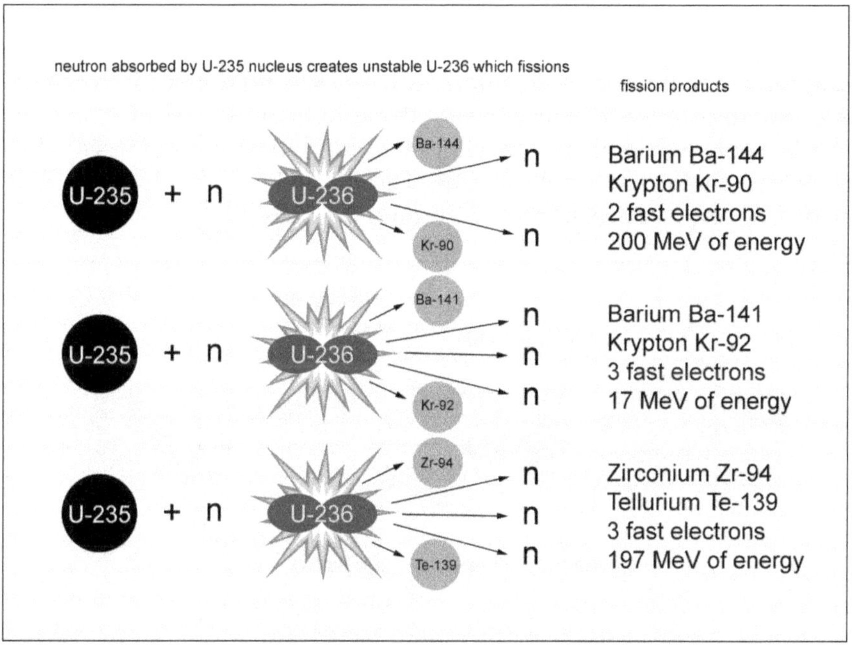

Possible pathways of U235 fission.

If there are other U-235 atoms close by they will be struck by the high-energy free neutrons escaping from the decay and will, in turn, be transformed into U-236 which will then decay to release more neutrons. This now becomes a chain reaction in which an ever-increasing number of U-235 atoms are split. If there is sufficient U-235 and the process is allowed to continue it will result in an explosion, releasing huge amounts of radioactive material and thermal energy. A small amount of U-235 cannot undergo a chain reaction because the chances of free neutrons striking other U-235 nuclei is small. However, if more U-235 is added, the chances of collision are increased until a 'critical mass' of U-235 is sufficient to start a chain reaction. A critical mass is the amount of a fissionable material that will support a self-sustaining chain reaction. The critical mass depends on the type of material: its purity, the temperature, the shape of the sample and how the neutron reactions are controlled.

U-235, which is found naturally in very small quantities in ores of U-238, undergoes fission more readily and emits more neutrons per fission than other such isotopes. The element Pu-239 has these same qualities. These are the primary fissionable materials used in atomic bombs.

In an atomic bomb, a mass of fissile material greater than the critical mass must be assembled instantaneously from two sub-critical masses and held together for about a millionth of a second to permit the chain reaction to propagate before the bomb explodes. One way of achieving this is using high explosives to shoot two subcritical slugs of fissionable material together in a hollow tube. The 'Little Boy' atomic bomb that was dropped on Hiroshima was triggered in this way.

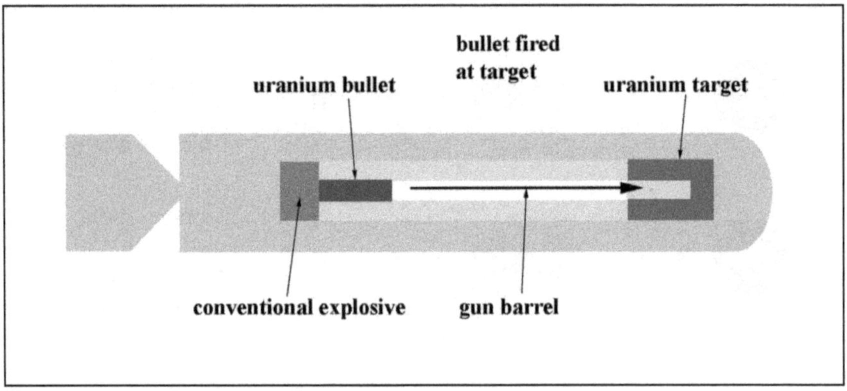

Gun mechanism for 'Little Boy' atomic bomb.

A second method, which was used in the second bomb ('Fat Man' that was dropped on Nagasaki is to suddenly implode a core of fissionable material and compress it into a smaller size and thus a greater density which has the effect of creating a critical mass. The core of an implosion-type atomic bomb consists of a sphere or a series of concentric shells of fissionable material surrounded by a jacket of high explosives, which, being simultaneously detonated, implode the fissionable material under enormous pressures into a denser mass that immediately achieves criticality.

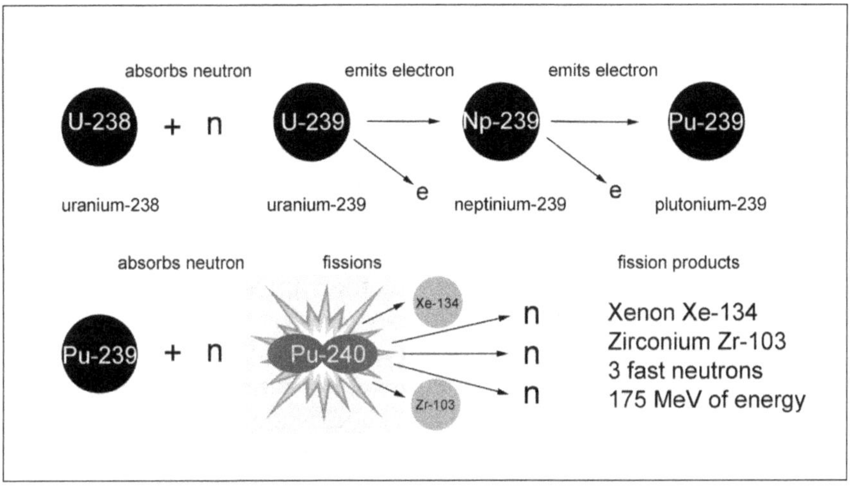

Plutonium-239 fission.

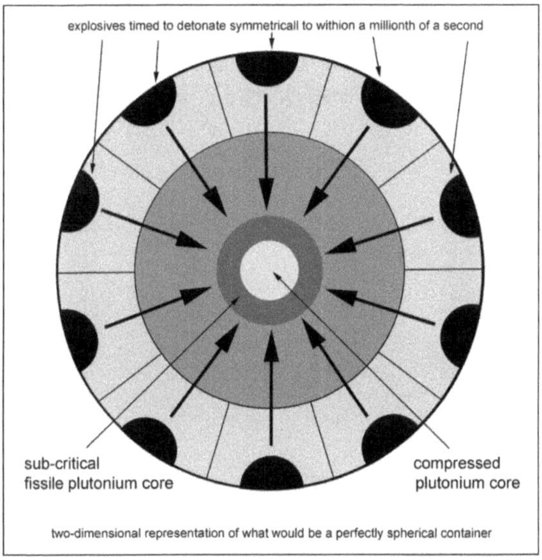

Implosion mechanism for 'Fat Man'
atomic bomb.

The high explosive surrounding the fissile material is ignited. A compressional shock wave begins to move inward. This creates a large increase in pressure and impinges on all points on the surface of the sphere of the fissile material in the bomb core at the same instant. As the core density increases, the mass becomes critical and then supercritical. Release of the initiator now produces large numbers of neutrons and the chain reaction continues until the energy generated inside the bomb becomes so great that the internal pressure due to the energy of the fission fragments exceeds the implosion pressure due to the shock wave and the bomb detonates

The detonation of an atomic bomb releases enormous amounts of thermal energy, achieving temperatures of several million degrees in the exploding bomb itself as well as a shockwave that propagates out from the blast to a distance of several miles. Large quantities of lethal radiation are also released. The 'Little Boy' bomb contained about 50kg of U-235 and had the explosive power of 15,000 tons of TNT. If all of the U-235 fuel had been used, the yield would have been 500 kilotons but the actual yield was only 3 per cent of that figure because, as the weapon exploded, most of the U-235 was dispersed before it could capture a neutron and contribute to the chain reaction.

Appendix 2

OPERATION FRESHMAN CASUALTIES

Bailey, Ernst William, aged 31
Belfield, John Thomas Vernon, aged 26
Bevan, Howell, aged 22
Blackburn, James Frank, aged 28
Bonner, Frank, aged 25
Bray, Frederick William, aged 29
Buckton, Albert, aged 23
Cairncross, James Dobson, aged 22
Campbell, Alexander, aged 22
Davies, Norman Arthur
de Gency, Gerard Walter Sewell, aged 20
Doig, Peter, aged 25
Edwards, George Mercier, aged 24
Falcone, James, aged 20
Farrell, Peter Paul, aged 26
Fraser, Herbert John, aged 28
Grundy, Charles Henry, aged 22
Haward, Arnold Thomas, aged 28
Healey, Fredrick, aged 29
Hunter, John Glen Vernon, aged 22
Jackon, Wallis Mahlon, aged 21
Jacques, William, aged 24
Knowles, George, aged 28
Legate, Herbert J, age not known
Masters, Trevor Louis, aged 25
Methven, David Alexander, aged 20
Norman, Robert, aged 22

Parkinson, Arthur Roland, aged 26
Pendlebury, Ernst, aged 25
Simkins, George, aged 30
Smallman, Leslie, age not known
Smith, Eric John, aged 24
Stephen, James May, age not known
Strathdee, Malcolm Frederick, age not known
Thomas, Arthur Edwin, aged 32
Thomas, John George Llewllyn, aged 23
Walsh, John Wilfred, aged 26
White, Thomas, aged 22
Williams, Gerald Stanley, aged 18

NOTES

Chapter 1: The Structure of Matter

1. Vince Houghton, *The Nuclear Spies* (Cornell University Press, 2019), p. 10.

Chapter 2: Splitting the Atom

1. Céline Jurgensen and Dominique Mongin, *France and Nuclear Deterrence; A Spirit of Resistance* (Fondation pour la Recherche Stratégique, 2020), p. 16.
2. Sam Kean, *The Bastard Brigade* (Hodder and Stoughton, 2019), p. 38.
3. Houghton, p. 12.
4. Andrew Norman, *The Amazing Story of Lisa Meitner* (Pen and Sword, 2021), p. 45.
5. Melanie Fine, *Fission Girl: Lise Meitner's Escape from Nazi Germany and Her Role in the Manhattan Project* (Kindle edition, 2017), p. 10.
6. Richard Rhodes, *The Making of the Atomic Bomb* (Simon and Schuster, 2012), p. 251.
7. Norman, p. 53.
8. Fine, p. 61.
9. Norman, p. 54.
10. Thomas Powers, *Heisenberg's War* (Alfred A Knopf, 1993), p. 111.
11. Norman, p. 60.
12. Fine, p. 68.
13. Ibid., p. 69.
14. Rhodes, p. 250.
15. Ibid., p. 253.
16. Ruth Lewin Sime, *Lise Meitner and the discovery of fission* (Sacramento City College, 2015).
17. Rhodes, p. 262.
18. Lise Meitner and Otto Frisch, 'Disintegration of Uranium by Neutrons: A New Type of Nuclear Reaction', *Nature* 143 (1939), p. 143.
19. Fine, pp. 86–7.

20. Fine, p. 85.
21. Norman, p. 97.
22. Rhodes, p. 261.
23. Paul Lawrence Rose, *Heisenberg and the Nazi Atomic Bomb Project* (University of California Press, 1998), pp. 83–4.
24. Al Cimino, *The Manhattan Project* (Arcturus Publishing, 2015), p. 24.
25. Rose, p. 88.
26. Ibid., p. 89.
27. Horst Kant, *Werner Heisenberg and the German Uranium Project* (Max Planck Institute for the History of Science, 2002), p. 3.
28. Ken Wilber, *Quantum Questions: Mystical Writings of the World's Great Physicists* (Kindle edition, 2001).
29. Jim Baggott, *Atomic; The First War of Physics and the Secret History of the Atom Bomb 1939-1949* (Icon Books, 2009), p. 16.
30. Kean, p. 73.
31. Baggott, p. 7.
32. Ibid., p. 36.
33. Rose, p. 125.
34. Fine, p. 89.
35. Cynthia C. Kelly, *The Manhattan Project* (Black Dog and Leventhal, 2009), p. 24.
36. Ibid., p. 26.
37. Fine, p. 92.
38. Nigel West, *Mortal Crimes* (Enigma Books, 2004), p. 28.
39. Kelly, p. 48.
40. Carl H. Meyer and Günter Schwartz, *The Theory of Nuclear Explosives That Heisenberg Did Not Present to the German Military* (Max Planck Institute for the History of Science, 2015), p. 15.
41. Kant, p. 8.
42. Rose, p. 109.
43. Mark Walker, *Nazi Science: Myth, Truth and the German Atomic Bomb* (Plenum Press, 1995), p. 36.
44. Rose, p. 112.
45. Ibid., p. 115.

Chapter 3: The Frisch–Peierls Memorandum

1. Graham Farmelo, *Churchill's Bomb* (Faber and Faber 2014, Kindle edition).
2. Norman Ridley, *The Role of Intelligence in the Battle of Britain* (Pen and Sword, 2021), p. 94.
3. Kelly, p. 32.

Chapter 4: Heavy Water

1. Ole Kristian Grimnes, *The Allied heavy water operations at Rjukan* (Norwegian Institute for Defence Studies, 1995), p. 8.
2. Bertrand Goldschmidt, *The supplies of Norwegian heavy water to France and the early development of atomic energy* (Norwegian Institute for Defence Studies, 1995), p. 19.
3. Jurgensen and Mongin, p. 27.
4. Kean, p. 107.
5. West, p. 6.
6. Baggott, p. 34.
7. Kerin Freeman, *The Civilian Bomb-Disposing Earl* (Pen and Sword, 2015), p. 84.
8. Jurgensen and Mongin, p. 29.
9. Freeman, p. 17.
10. P.M.S. Blackett, *Jean Frédéric Joliot. 1900-1958* (Biographical Memoirs of Fellows of the Royal Society, 1960), p. 96.

Chapter 5: Tube Alloys

1. Rose, p. 99.
2. West, p. 6.
3. Farmelo.
4. Ibid.
5. West, p. 19.
6. Charmian Brinson and Richard Dove, *A Matter of Intelligence* (Manchester University Press, 2014), p. 196.
7. Nancy Thorndike Greenspan, *Atomic Spy; The Dark Lives of Klaus Fuchs* (Penguin, 2020), p. 88.
8. Ibid., p. 91.
9. Ibid., pp. 106–07.
10. Ibid., p. 112.
11. Ibid., p. 123.
12. Brinson and Dove, p. 197.
13. Greenspan, p. 135.
14. M.A.U.D. Committee, Ministry of Aircraft Production, London, July 1941.
15. Rhodes, p. 372.
16. Farmelo.
17. Brinson and Dove, p. 116.
18. West, p. 57.
19. Brinson and Dove, p. 200.
20. Greenspan, p. 140.

Chapter 6: The US Office of Scientific Research

1. Jones, p. 13.
2. Baggott, p. 30.
3. Rose, p. 86.
4. Rhodes, p. 336.
5. Ibid., p. 359.
6. Ibid., p. 367.
7. Ibid., p. 368.
8. Jones, p. 27.
9. Baggott, p. 109.
10. Farmelo.
11. Kelly, p. 44.
12. Baggot, p. 112.
13. Farmelo.
14. Susan Williams, *Spies in the Congo* (Hurst and Company, 2016), p. 32.
15. Ibid., p. 35.

Chapter 7: Werner Heisenberg

1. Powers, p. 7.
2. Ibid., p. 4.
3. Rose, p. xv.
4. Baggott, p. 39.
5. Rhodes, p. 346.
6. Walter E. Grunden, Yutaka Kawamura, Eduard Kolchinsky, Helmut Maier and Masakatsu Yamazaki, *Laying the Foundation for Wartime Research: A Comparative Overview of Science Mobilisation in National Socialist Germany, Japan and the Soviet Union* (The History of Science Society, 2005), p. 84.
7. Kean, p. 91.
8. Rose, p. 126.
9. Powers, p. 100.
10. Rose, p. 137.
11. Baggott, p. 82.
12. Rose, p. 145.
13. Fine, pp. 98–9.
14. Powers, p. 116.
15. Rose, p. 276.
16. Fine, p. 98.
17. Powers, p. 126.
18. Baggott, p. 119.

19. Powers, p. 131.
20. Rhodes, p. 403.
21. Powers, p. 133.
22. Hans Bethe, *The German Uranium Project* (American Institute of Physics, 2002).
23. Rhodes, p. 403
24. Bethe.
25. Powers, p. 148.
26. Rhodes, p. 405.
27. Powers, p. 150.
28. Rose, p. 185.
29. Powers, p. 161.
30. Ibid., p. 163.

Chapter 8: The Manhattan Project

1. Farmelo.
2. Stanley Goldberg, 'Groves and Oppenheimer: The Story of a Partnership', *The Antioch Review* Vol. 53, No. 4 (1995), p. 481.
3. Ibid., p. 484.
4. Kelly, p. 76.
5. Baggott, p. 164.
6. Rhodes, p. 502.
7. *The Conferences at Washington, 1941-1942 and Casablanca* (United States, Department of State, Historical Office 1943).
8. Williams, p. 9.
9. Houghton, p. 38.
10. Kelly, p. 77.
11. Alex Wellerstein, 'The curious death of Oppenheimer's mistress', *The Nuclear Secrecy Blog* (2015).
12. Baggott, p. 200.
13. Williams, p. 166.
14. Donald Howard Avery, 'Atomic Scientific Co-operation and Rivalry Among Allies: The Anglo-Canadian Montreal Laboratory and the Manhattan Project, 1943–1946', *War in History* Vol. 2, No. 3 (1995), p. 277.
15. Ibid., p. 283.
16. Ibid., p. 293.
17. Ibid., p. 286.
18. Ibid., p. 296.
19. Baggott, p. 397.

Chapter 9: Soviet Spies

1. Greenspan, p. 146.
2. David Holloway, *Stalin and the Bomb* (Yale University Press, 1994), p. 109.
3. Gregg Herken, 'Target Enormoz', *Journal of Cold War Studies* Vol. 11, No. 3 (2009), p. 71.
4. West, p. 196.
5. Herken, p. 74.
6. West, p. 69.
7. Herken, p. 80.
8. West, p. 194.
9. Ibid., p. 113.
10. Baggott, p. 195.
11. Wellerstein, 'The curious death of Oppenheimer's mistress'.
12. Goldberg, p. 492.
13. Kelly, p. 237.
14. Ibid., p. 239.

Chapter 10: Grouse, Freshman and Gunnerside

1. Baggott, p. 131.
2. Goodman, p. 104.
3. Baggott, p. 133.
4. Joachim Rønneberg, *Operation Gunnerside* (Norwegian Institute for Defence Studies, 1995).
5. Ibid.
6. Baggott, p. 233.

Chapter 11: The Soviet Atomic Bomb Project

1. David Holloway, *Stalin and the Bomb* (Yale University Press, 1994), p. 27.
2. Ibid., p. 32.
3. Lee G. Pondrom, *The Soviet atomic project: how the Soviet Union obtained the atomic bomb* (World Scientific Publishing Co. Pte. Ltd, 2018), p. 155.
4. Holloway, *Stalin and the Bomb*, p. 36.
5. Pondrom, p. 169.
6. Holloway, *Stalin and the Bomb*, p. 53.
7. David Holloway, 'Entering the Nuclear Arms Race: The Soviet Decision to Build the Atomic Bomb, 1939-45', *Social Studies of Science* Vol. 11, No. 2 (1981), p. 168.
8. Pondrom, p. 184.

9. Ibid., p. 191.
10. Ibid., p. 186.
11. Ibid., p. 179.
12. Brinson and Dove, p. 210.
13. Ibid., p. 214.
14. West, p. 50.
15. *Edith Tudor-Hart in the Shadow of Tyranny* (National Galleries of Scotland), p. 121.
16. Brinson and Dove, p. 83.
17. Ibid., p. 190.
18. Ibid.
19. Pondrom, p. 297.
20. Baggott, p. 249.
21. Holloway, *Stalin and the Bomb*, p. 103.
22. Ibid., p. 86.
23. Pondrom, p. 267.

Chapter 12: The Japanese Atomic Bomb

1. Grunden et al., p. 111.
2. Robert K. Wilcox, *Japan's Secret War* (Permuted Press, 2019), p. 56.
3. Grunden et al., p. 114.
4. Wilcox, p. 57.
5. Ibid., p. 58.
6. Ibid., p. 98.
7. Ibid., p. 138.

Chapter 13: The Alsos Missions

1. Jeffrey T. Richelson, *Spying on the Bomb* (Norton and Company, 2006, Kindle edition).
2. John V.H. Dippel, *Two Against Hitler; Stealing the Nazi's Best-Kept Secrets* (Praeger, 1992), p. 93.
3. Ibid., p. 84.
4. Ibid., p. 83.
5. Ibid., p. 91.
6. Richelson.
7. Houghton, p. 77.
8. Richelson.
9. Houghton, p. 105.
10. Kean, p. 319.
11. Richelson.
12. Kean, p. 313.

13. Martijn van Calmthout, *Sam Goudsmit and the Hunt for Hitler's Atom Bomb* (Prometheus Books, 2018), p. 78.
14. Richelson.
15. Kean, p. 312.
16. Richelson.
17. Houghton, p. 128.
18. Powers, p. 369.
19. Calmthout, p. 114.
20. Kelly, p. 259.
21. Powers, p. 425.
22. Baggott, p. 43.
23. Ibid., p. 350.

Chapter 14: 'Little Boy' and 'Fat Man'

1. Baggott, p. 306.
2. Holloway, *Stalin and the Bomb* (Yale University Press 1994), p. 130.
3. Baggott, p. 400.
4. Ibid., p. 401.
5. Ibid., p. 441.
6. Kelly, p. 249.

Chapter 15: 'Joe-1'

1. Pondrom, p. 263.
2. Jeffrey T. Richelson, *Spying on the Bomb* (Norton and Company 2006).
3. Baggott, p. 296.
4. Holloway, *Stalin and the Bomb*, p. 130.
5. Baggott, pp. 360–1.
6. Holloway, *Stalin and the Bomb*, p. 137.
7. Baggott, p. 382.
8. John Earl Haynes and Harvey Kleph, *Early Cold War Spies: The Espionage Trials that Shaped American Politics* (Cambridge University Press, 2006), p. 48.
9. Holloway, *Stalin and the Bomb*, p. 144.
10. Pondrom, p. 376.
11. Holloway, *Stalin and the Bomb*, p. 121.
12. Houghton, p. 150.
13. Ibid., p. 156.
14. Ibid., p. 177.
15. Ibid., p. 161.
16 Ibid., p. 164.
17. Holloway, *Stalin and the Bomb*, p. 187.

18. Pondrom, p. 473.
19. Ibid., p. 552.
20. Ibid., p. 550.
21. Ibid., p. 627.
22. Ibid., p. 638.
23. Ibid., pp. 639–40.
24. Houghton, p. 179.

SOURCES

Avery, Donald Howard, 'Atomic Scientific Co-operation and Rivalry Among Allies: The Anglo-Canadian Montreal Laboratory and the Manhattan Project, 1943–1946', *War in History* Vol. 2, No. 3 (1995).

Baggott, Jim, *Atomic; The First War of Physics and the Secret History of the Atom Bomb 1939-1949* (Icon Books, 2009).

Benke, Richard, *Japanese Uranium Oxide* (Associated Press, Los Alamos, 1997).

Bernstein, Barton J., 'The Uneasy Alliance: Roosevelt, Churchill and the Atomic Bomb, 1940-1945', *The Western Political Quarterly* Vol. 29, No. 2 (1976).

Bethe, Hans, *The German Uranium Project* (American Institute of Physics, 2002).

Blackett, P.M.S., *Jean Frédéric Joliot. 1900-1958* (Biographical Memoirs of Fellows of the Royal Society, 1960).

Brinson, Charmian and Dove, Richard, *A Matter of Intelligence* (Manchester University Press, 2014).

Calmthout, Martijn van, *Sam Goudsmit and the Hunt for Hitler's Atom Bomb* (Prometheus Books, 2018).

Cimino, Al, *The Manhattan Project* (Arcturus Publishing, 2015).

Dippel, John V.H., *Two Against Hitler; Stealing the Nazi's Best-Kept Secrets* (Praeger, 1992).

Easley, Matt, *The Atomic Bomb That Never Was: Germany's Atomic Bomb Project* (vanderbilt.edu).

Edith Tudor-Hart in the Shadow of Tyranny (National Galleries of Scotland).

Farmelo, Graham, *Churchill's Bomb* (Faber and Faber, 2014, Kindle edition).

Fine, Melanie, *Fission Girl* (Kindle edition, 2017).

Freeman, Kerin, *The Civilian Bomb-Disposing Earl* (Pen and Sword, 2015).

Goldberg, Stanley, 'Groves and Oppenheimer: The Story of a Partnership', *The Antioch Review* Vol. 53, No. 4 (1995).

Goldschmidt, Bertrand, *How it all began in Canada; The role of French scientists* (Canadian Nuclear Society).

__________________, *The supplies of Norwegian heavy water to France and the early development of atomic energy* (Norwegian Institute for Defence Studies, 1995).

Goodman, Michael S., 'MI6's Atomic Man', *War in History* Vol. 23, No. 1 (2016).

Greenspan, Nancy Thorndike, *Atomic Spy; The Dark Lives of Klaus Fuchs* (Penguin, 2020).

Grimnes, Ole Kristian, *The Allied heavy water operations at Rjukan* (Norwegian Institute for Defence Studies, 1995).

Grunden, Walter E., Kawamura, Yutaka, Kolchinsky, Eduard, Maier Helmut and Yamazaki, Masakatsu, *Laying the Foundation for Wartime Research: A Comparative Overview of Science Mobilisation in National Socialist Germany, Japan and the Soviet Union* (The History of Science Society, 2005).

________________, Walker, Mark and Yamazaki Masakatsu, 'Wartime Nuclear Weapons Research in Germany and Japan', *Osiris* 2nd Series, Vol. 20 (2005).

Haynes, John Earl and Kleph, Harvey, *Early Cold War Spies: The Espionage Trials that Shaped American Politics* (Cambridge University Press, 2006).

Herken, Gregg, 'Target Enormoz', *Journal of Cold War Studies* Vol. 11, No. 3 (2009).

Holloway, David, 'Entering the Nuclear Arms Race: The Soviet Decision to Build the Atomic Bomb, 1939-45', *Social Studies of Science* Vol. 11, No. 2 (1981).

______________, *Stalin and the Bomb* (Yale University Press, 1994).

Houghton, Vince, *The Nuclear Spies* (Cornell University Press, 2019).

Hyde, H. Montgomery, *The Atom Bomb Spies* (Hamish Hamilton, 1980).

Jones, Vincent C., *The Army and the Atomic Bomb*, Volume 8, Part 9 (Centre of Military History; United States Army, 1985).

Jurgensen, Céline and Mongin, Dominique, *France and Nuclear Deterrence; A Spirit of Resistance* (Fondation pour la Recherche Stratégique, 2020).

Kant, Horst, *Werner Heisenberg and the German Uranium Project* (Max Planck Institute for the History of Science, 2002).

Kean, Sam, *The Bastard Brigade* (Hodder and Stoughton, 2019).

Kelly, Cynthia C., *The Manhattan Project* (Black Dog and Leventhal, 2009).

Lokhova, Svetlana, *The Spy who Changed History* (William Collins, 2018).

Macintyre, Ben, *Agent Sonya* (Penguin, 2021).

M.A.U.D. Committee (Ministry of Aircraft Production, London, 1941).

Meitner, Lise and Frisch, Otto, 'Disintegration of Uranium by Neutrons: A New Type of Nuclear Reaction', *Nature* 143 (1939).

Meyer, Carl H. and Schwartz, Günter, *The Theory of Nuclear Explosives That Heisenberg Did Not Present to the German Military* (Max Planck Institute for the History of Science, 2015).

Moss, Norman, *Klaus Fuchs* (Grafton Books, 2018).

Neff, Robert, 'Japan Tested Atomic Bomb in NK Before End of WWII?', *The Korea Times*, 4 December 2009.

Norman, Andrew, *The Amazing Story of Lisa Meitner* (Pen and Sword, 2021).

Oleynikov, Pavel V., 'Germany's Atomic Bomb Project', *The Nonproliferation Review* (2000).

Pondrom, Lee G., *The Soviet atomic project : how the Soviet Union obtained the atomic bomb* (World Scientific Publishing Co. Pte. Ltd., 2018).

Powers, Thomas, *Heisenberg's War* (Alfred A. Knopf, 1993).

Rhodes, Richard, *The Making of the Atomic Bomb* (Simon and Schuster, 2012).

Richelson, Jeffrey T., *Spying on the Bomb* (Norton and Company, 2006 Kindle edition).

Ridley, Norman, *The Role of Intelligence in the Battle of Britain* (Pen and Sword, 2021).

Rønneberg, Joachim, *Operation Gunnerside* (Norwegian Institute for Defence Studies, 1995).

Rose, Paul Lawrence, *Heisenberg and the Nazi Atomic Bomb Project* (University of California Press, 1998).

Sime, Ruth Lewin, *Lise Meitner and the discovery of fission* (Sacramento City College, 2015).

Smyth, H. D., *Atomic Energy for Military Purposes (The Smyth Report)* (1945).

The Conferences at Washington, 1941-1942 and Casablanca (United States. Department of State. Historical Office, 1943).

Walker, Mark, *Nazi Science: Myth, Truth and the German Atomic Bomb* (Plenum Press, 1995).

Wellerstein, Alex, *Bulletin of Atomic Scientists* (2008).

_____________, 'The curious death of Oppenheimer's mistress', *The Nuclear Secrecy Blog* (2015).

West, Nigel, *Mortal Crimes* (Enigma Books, 2004).

Wilber, Ken, *Quantum Questions: Mystical Writings of the World's Great Physicists* (Kindle edition, 2001).

Wilcox, Robert K., *Japan's Secret War* (Permuted Press, 2019).

Williams, Susan, *Spies in the Congo* (Hurst and Company, 2016).

INDEX